Computational Mathematics, Algorithms, and Data Processing

Computational Mathematics, Algorithms, and Data Processing

Editors

Daniele Mortari
Yalchin Efendiev
Boris Hanin

MDPI • Basel • Beijing • Wuhan • Barcelona • Belgrade • Manchester • Tokyo • Cluj • Tianjin

Editors
Daniele Mortari
Department of Aerospace
 Engineering, Texas A&M
University—3141 TAMU
College Station
USA

Yalchin Efendiev
Mobil Chair in Computational Sciences,
Professor of Mathematics,
Director of Institute for
Scientific Computation (ISC),
Department of Mathematics & ISC,
Texas A&M University
USA

Boris Hanin
Department of Mathematics,
Texas A&M University
USA

Editorial Office
MDPI
St. Alban-Anlage 66
4052 Basel, Switzerland

This is a reprint of articles from the Special Issue published online in the open access journal *Mathematics* (ISSN 2227-7390) (available at: https://www.mdpi.com/journal/mathematics/special_issues/Computational_Mathematics_Algorithms_Data_Processing).

For citation purposes, cite each article independently as indicated on the article page online and as indicated below:

LastName, A.A.; LastName, B.B.; LastName, C.C. Article Title. *Journal Name* **Year**, *Article Number*, Page Range.

ISBN 978-3-03943-591-3 (Hbk)
ISBN 978-3-03943-592-0 (PDF)

© 2020 by the authors. Articles in this book are Open Access and distributed under the Creative Commons Attribution (CC BY) license, which allows users to download, copy and build upon published articles, as long as the author and publisher are properly credited, which ensures maximum dissemination and a wider impact of our publications.

The book as a whole is distributed by MDPI under the terms and conditions of the Creative Commons license CC BY-NC-ND.

Contents

About the Editors vii

Preface to "Computational Mathematics, Algorithms, and Data Processing" ix

Daniele Mortari and Carl Leake
The Multivariate Theory of Connections †
Reprinted from: *Mathematics* **2019**, *7*, 296, doi:10.3390/math7030296 1

Zecheng Zhang, Eric T. Chung, Yalchin Efendiev and Wing Tat Leung
Learning Algorithms for Coarsening Uncertainty Space and Applications to Multiscale Simulations
Reprinted from: *Mathematics* **2020**, *8*, 720, doi:10.3390/math8050720 23

Christine Schmid and Kyle J. DeMars
Angular Correlation Using Rogers-Szegő-Chaos
Reprinted from: *Mathematics* **2020**, *8*, 171, doi:10.3390/math8020171 41

Le Zou, Liangtu Song, Xiaofeng Wang, Yanping Chen, Chen Zhang and Chao Tang
Bivariate Thiele-Like Rational Interpolation Continued Fractions with Parameters Based on Virtual Points
Reprinted from: *Mathematics* **2020**, *8*, 71, doi:10.3390/math8010071 65

Denis Spiridonov, Jian Huang, Maria Vasilyeva, Yunqing Huang and Eric T. Chung
Mixed Generalized Multiscale Finite Element Method for Darcy-Forchheimer Model
Reprinted from: *Mathematics* **2019**, *7*, 1212, doi:10.3390/math7121212 87

Qiuyan Xu and Zhiyong Liu
Scattered Data Interpolation and Approximationwith Truncated Exponential Radial Basis Function
Reprinted from: *Mathematics* **2019**, *7*, 1101, doi:10.3390/math7111101 101

Boris Hanin
Universal Function Approximation by Deep Neural Nets withBounded Width and ReLU Activations
Reprinted from: *Mathematics* **2019**, *7*, 992, doi:10.3390/math7100992 115

Min Wang, Siu Wun Cheung, Eric T. Chung, Yalchin Efendiev, Wing Tat Leung and Yating Wang
Prediction of Discretization of GMsFEM Using Deep Learning
Reprinted from: *Mathematics* **2019**, *7*, 412, doi:10.3390/math7050412 125

Changbum Chun and Beny Neta
Trigonometrically-Fitted Methods: A Review
Reprinted from: *Mathematics* **2019**, *7*, 1197, doi:10.3390/math7121197 141

About the Editors

Daniele Mortari is a Professor of Aerospace Engineering at Texas A&M University, working on the field of attitude and position estimation, satellite constellation design, sensor data processing, and various topic in linear algebra and numerical algorithms. In addition, he has taught at the School of Aerospace Engineering of Rome's University, and at Electronic Engineering of Perugia's University, both in Italy. He received his dottore degree in Nuclear Engineering from the University of Rome "La Sapienza," in 1981. He has published more than 300 papers, holds a U.S. patent, and has been widely recognized for his work, including receiving best paper Award from AAS/AIAA, three NASA's Group Achievement Awards, 2003 Spacecraft Technology Center Award, 2007 IEEE Judith A. Resnik Award, and 2016 AAS Dirk Brouwer Award. He is a member of the International Academy of Astronautics, an IEEE and AAS Fellow, an AIAA Associate Fellow, an Honorary Member of IEEE-AESS Space System Technical Panel, and a former IEEE Distinguish Speaker.

Yalchin Efendiev is a professor in Mathematics at Texas A&M University. He works on multiscale methods, learning for multiscale problems, and uncertainty quantification.

Boris Hanin is an assistant professor in Operations Research and Financial Engineering at Princeton University. He works on probability, deep learning, and mathematical physics. Before joining the faculty at Princeton, he was an assistant professor in Mathematics at Texas A&M and an NSF postdoctoral fellow in Mathematics at MIT. He has also held visiting positions at Facebook AI Research, Google, and the Simons Institute for the Theory of Computation.

Preface to "Computational Mathematics, Algorithms, and Data Processing"

This Special Issue "Computational Mathematics, Algorithms, and Data Processing" of MDPI consists of original articles on new mathematical tools and numerical methods for computational problems. This Special Issue of MDPI is motivated by the recent profusion and success of large-scale numerical methods in a variety of applied problems and is focused specifically on ideas that are scalable to large scale problems and have the potential the significantly improve the current state of art practices. The topics covered include: numerical stability, interpolation, approximation, complexity, numerical linear algebra, differential equations (ordinary, partial), optimization, integral equations, systems of nonlinear equations, compression or distillation, and active learning. All articles include a discussion of theoretical guarantees or at least justifications for the methods.

Daniele Mortari, Yalchin Efendiev, Boris Hanin
Editors

Article

The Multivariate Theory of Connections [†]

Daniele Mortari * and Carl Leake *

Aerospace Engineering, Texas A&M University, College Station, TX 77843, USA
* Correspondence: mortari@tamu.edu (D.M.); leakec@tamu.edu (C.L.); Tel.: +1-979-845-0734 (D.M.)
† This paper is an extended version of our paper published in Mortari, D. "The Theory of Connections: Connecting Functions." IAA-AAS-SciTech-072, Forum 2018, Peoples' Friendship University of Russia, Moscow, Russia, 13–15 November 2018.

Received: 4 January 2019; Accepted: 18 March 2019; Published: 22 March 2019

Abstract: This paper extends the univariate Theory of Connections, introduced in (Mortari, 2017), to the multivariate case on rectangular domains with detailed attention to the bivariate case. In particular, it generalizes the bivariate Coons surface, introduced by (Coons, 1984), by providing analytical expressions, called *constrained expressions*, representing *all* possible surfaces with assigned boundary constraints in terms of functions and arbitrary-order derivatives. In two dimensions, these expressions, which contain a freely chosen function, $g(x,y)$, satisfy all constraints no matter what the $g(x,y)$ is. The boundary constraints considered in this article are Dirichlet, Neumann, and any combinations of them. Although the focus of this article is on two-dimensional spaces, the final section introduces the *Multivariate Theory of Connections*, validated by mathematical proof. This represents the multivariate extension of the Theory of Connections subject to arbitrary-order derivative constraints in rectangular domains. The main task of this paper is to provide an analytical procedure to obtain constrained expressions in any space that can be used to transform constrained problems into unconstrained problems. This theory is proposed mainly to better solve PDE and stochastic differential equations.

Keywords: interpolation; constraints; embedded constraints

1. Introduction

The Theory of Connections (ToC), as introduced in [1], consists of a general analytical framework to obtain *constrained expressions*, $f(x)$, in one-dimension. A constrained expression is a function expressed in terms of another function, $g(x)$, that is freely chosen and, no matter what the $g(x)$ is, the resulting expression always satisfies a set of n constraints. ToC generalizes the one-dimensional interpolation problem subject to n constraints using the general form,

$$f(x) = g(x) + \sum_{k=1}^{n} \eta_k \, p_k(x), \tag{1}$$

where $p_k(x)$ are n user-selected linearly independent functions, η_k are derived by imposing the n constraints, and $g(x)$ is a *freely chosen* function subject to be *defined and nonsingular* where the constraints are specified. Besides this requirement, $g(x)$ can be any function, including, discontinuous functions, delta functions, and even functions that are undefined in some domains. Once the η_k coefficients have been derived, then Equation (1) satisfies all the n constraints, *no matter what the $g(x)$ function is*.

Constrained expressions in the form given in Equation (1) are provided for a wide class of constraints, including constraints on points and derivatives, linear combinations of constraints, as well as infinite and integral constraints [2]. In addition, weighted constraints [3] and point constraints on continuous and discontinuous periodic functions with assigned period can also be obtained [1]. How to extend ToC to inequality and nonlinear constraints is currently a work in progress.

The Theory of Connections framework can be considered the generalization of interpolation; rather than providing a class of functions (e.g., monomials) satisfying a set of n constraints, it derives *all* possible functions satisfying the n constraints by spanning all possible $g(x)$ functions. This has been proved in Ref. [1]. A simple example of a constrained expression is,

$$f(x) = g(x) + \frac{x(2x_2 - x)}{2(x_2 - x_1)} [\dot{y}_1 - \dot{g}(x_1)] + \frac{x(x - 2x_1)}{2(x_2 - x_1)} [\dot{y}_2 - \dot{g}(x_2)]. \qquad (2)$$

This equation always satisfies $\left.\frac{df}{dx}\right|_{x_1} = \dot{y}_1$ and $\left.\frac{df}{dx}\right|_{x_2} = \dot{y}_2$, as long as $\dot{g}(x_1)$ and $\dot{g}(x_2)$ are defined and nonsingular. In other words, *the constraints are embedded into the constrained expression*.

Constrained expressions can be used to transform constrained optimization problems into unconstrained optimization problems. Using this approach, fast least-squares solutions of linear [4] and nonlinear [5] ODE have been obtained at machine error accuracy and with low (actually, very low) condition number. Direct comparisons of ToC versus MATLAB's ode45 [6] and Chebfun [7] have been performed on a small test of ODE with excellent results [4,5]. In particular, the ToC approach to solve ODE consists of a unified framework to solve IVP, BVP, and multi-value problems. The extension of differential equations subject to component constraints [8] has opened the possibility for ToC to solve *in real-time* a class of direct optimal control problems [9], where the constraints connect state and costate.

This study first extends the Theory of Connections to two-dimensions by providing, for rectangular domains, *all* surfaces that are subject to: (1) Dirichlet constraints; (2) Neumann constraints; and (3) any combination of Dirichlet and Neumann constraints. This theory is then generalized to the Multivariate Theory of Connections which provide in n-dimensional space all possible manifolds that satisfy boundary constraints on the value and boundary constraints on any-order derivative.

This article is structured as follows. First, it shows that the one-dimensional ToC can be used in two dimensions when the constraints (functions or derivatives) are provided along one axis only. This is a particular case, where the original univariate theory [1] can be applied with basically no modifications. Then, a two dimensional ToC version is developed for Dirichlet type boundary constraints. This theory is then extended to include Neumann and mixed type boundary constraints. Finally, the theory is extended to n-dimensions and to incorporate arbitrary-order derivative boundary constraints followed by a mathematical proof validating it.

2. Manifold Constraints in One Axis, Only

Consider the function, $f(x)$, where $f : \mathbb{R}^n \to \mathbb{R}^1$, subject to one constraint manifold along the ith variable, x_i, that is, $f(x)|_{x_i=v} = c(x_i^v)$. For instance, in 3-D space, this can be the surface constraint, $f(x, y, z)|_{y=\pi} = c(x, \pi, z)$. *All manifolds* satisfying this constraint can be expressed using the additive form provided in Ref. [1],

$$f(x) = g(x) + [c(x_i^v) - g(x_i^v)]$$

where $g(x)$ is a freely chosen function that must be defined and nonsingular at the constraint coordinates. When m manifold constraints are defined along the x_i-axis, then the 1-D methodology [1] can be applied as it is. For instance, the constrained expression subject to m constraints along the x_i variable evaluated at $x_i = w_k$, where $k \in [1, m]$, that is, $f(x)|_{x_i=w_k} = c(x_i^{w_k})$, is,

$$f(x) = g(x) + \sum_{k=1}^{m} \left\{ [c(x_i^{w_k}) - g(x_i^{w_k})] \prod_{j \neq k} \frac{x_i - w_j}{w_k - w_j} \right\}. \qquad (3)$$

Note that this equation coincides with the Waring interpolation form (better known as Lagrangian interpolation form) [10] if the free function vanishes, $g(x) = 0$.

2.1. Example #1: Surface Subject to Four Function Constraints

The first example is designed to show how to use Equation (3) with mixed, continuous, discontinuous, and multiple constraints. Consider the following four constraints,

$$c(x,-2) = \sin(2x), \quad c(x,0) = 3\cos x\,[(x+1)\bmod(2)], \quad c(x,1) = 9e^{-x^2}, \quad \text{and} \quad c(x,3) = 1-x.$$

This example highlights that the constraints and free-function may be discontinuous by using the modular arithmetic function. The result is a surface that is continuous in x at some coordinates (at $y = -2, 1,$ and 3) and discontinuous at $y = 0$. The surfaces shown in Figures 1 and 2 were obtained using two distinct expressions for the free function, $g(x,y)$.

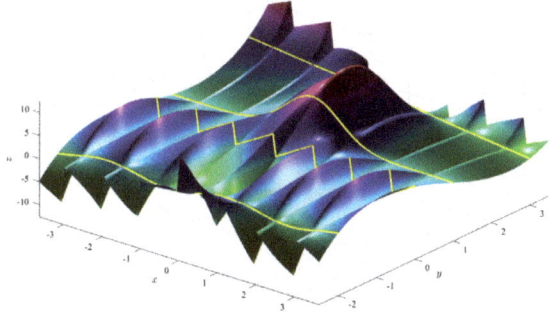

Figure 1. Surface obtained using function $g(x,y) = 0$ (simplest surface).

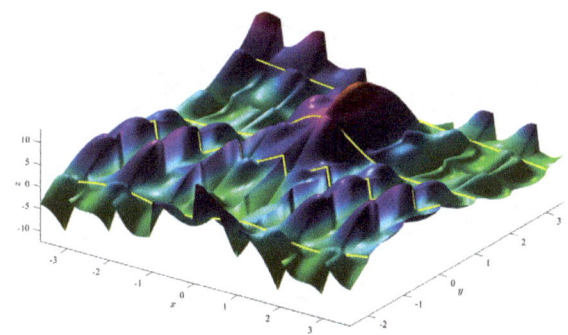

Figure 2. Surface obtained using function $g(x,y) = x^2 y - \sin(5x)\cos(4\bmod(y,1))$.

2.2. Example #2: Surface Subject to Two Functions and One Derivative Constraint

This second example is provided to show how to use the general approach given in Equation (1) and described in [1], when derivative constraints are involved. Consider the following three constraints,

$$c(x,-2) = \sin(2x), \quad c_y(x,0) = 0, \quad \text{and} \quad c(x,1) = 9e^{-x^2}.$$

Using the functions $p_1(y) = 1$, $p_2(y) = y$, and $p_3(y) = y^2$, the constrained expression form satisfying these three constraints assumes the form,

$$f(x,y) = g(x,y) + \eta_1(x) + \eta_2(x)\,y + \eta_3(x)\,y^2. \tag{4}$$

The three constraints imply the constraints,

$$\sin(2x) = g(x,-2) + \eta_1 - 2\eta_2 + 4\eta_3$$
$$0 = g_y(x,0) + \eta_2$$
$$9e^{-x^2} = g(x,1) + \eta_1 + \eta_2 + \eta_3,$$

from which the values of the η_k coefficients,

$$\eta_1 = 2g_y(x,0) + 12e^{-x^2} - \frac{\sin(2x)}{3} + \frac{1}{3}g(x,-2) - \frac{4}{3}g(x,1)$$
$$\eta_2 = -g_y(x,0)$$
$$\eta_3 = \frac{\sin(2x)}{3} - \frac{1}{3}g(x,-2) - g_y(x,0) - 3e^{-x^2} + \frac{1}{3}g(x,1),$$

can be derived. After substituting these coefficients into Equation (4), the constrained expression that always satisfies the three initial constraints is obtained. Using this expression and two different free functions, $g(x,y)$, we obtained the surfaces shown in Figures 3 and 4, respectively. The constraint $c_y(x,0) = 0$, difficult to see in both figures, can be verified analytically.

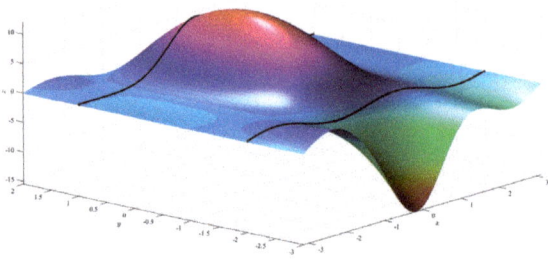

Figure 3. Surface obtained using function $g(x,y) = 0$ (simplest surface).

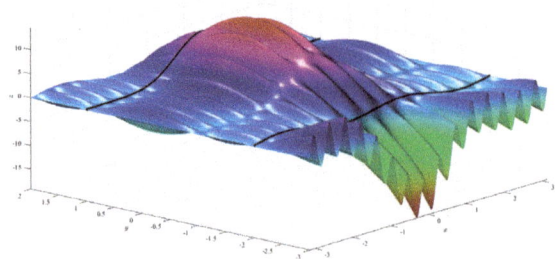

Figure 4. Surface obtained using function $g(x,y) = 3x^2y - 2\sin(15x)\cos(2y)$.

3. Connecting Functions in Two Directions

In this section, the Theory of Connections is extended to the two-dimensional case. Note that dealing with constraints in two (or more) directions (functions or derivatives) requires particular attention. In fact, two orthogonal constraint functions cannot be completely distinct as they intersect at one point where they need to match in value. In addition, if the formalism derived for the 1-D case is applied to 2-D case, some complications arise. These complications are highlighted in the following simple clarifying example.

Consider the two boundary constraint functions, $f(x,0) = q(x)$ and $f(0,y) = h(y)$. Searching the constrained expression as originally done for the one-dimensional case implies the expression,

$$f(x,y) = g(x,y) + \eta_1 \, p_1(x,y) + \eta_2 \, p_2(x,y).$$

The constraints imply the two constraints,

$$\begin{cases} q(x) = g(x,0) + \eta_1 \, p_1(x,0) + \eta_2 \, p_2(x,0) \\ h(y) = g(0,y) + \eta_1 \, p_1(0,y) + \eta_2 \, p_2(0,y). \end{cases}$$

To obtain the values of η_1 and η_2, the determinant of the matrix to invert is $p_1(x,0) \, p_2(0,y) - p_1(0,y) \, p_2(x,0)$. This determinant is y by selecting $p_1(x,y) = 1$ and $p_2(x,y) = y$, or it is x by selecting $p_1(x,y) = x$ and $p_2(x,y) = 1$. Therefore, to avoid singularities, this approach requires paying particular attention to the domain definition and/or on the user-selected functions, $p_k(x,y)$. To avoid dealing with these issues, a new (equivalent) formalism to derive constrained expressions is devised for the higher dimensional case.

The Theory of Connections extension to the higher dimensional case (with constraints on all axes) can be obtained by re-writing the constrained expression into an equivalent form, highlighting a general and interesting property. Let us show this by an example. Equation (2) can be re-written as,

$$f(x) = \underbrace{\frac{x(2x_2 - x)}{2(x_2 - x_1)} \dot{y}_1 + \frac{x(x - 2x_1)}{2(x_2 - x_1)} \dot{y}_2}_{A(x)} + \underbrace{g(x) - \frac{x(2x_2 - x)}{2(x_2 - x_1)} \dot{g}_1 - \frac{x(x - 2x_1)}{2(x_2 - x_1)} \dot{g}_2}_{B(x)}. \qquad (5)$$

These two components, $A(x)$ and $B(x)$, of a constrained expression have a specific general meaning. The term, $A(x)$, represents an (*any*) interpolating function satisfying the constraints while the $B(x)$ term represents *all* interpolating functions that are vanishing at the constraints. Therefore, the generation of all functions satisfying multiple orthogonal constraints in n-dimensional space can always be expressed by the general form, $f(x) = A(x) + B(x)$, where $A(x)$ is *any* function satisfying the constraints and $B(x)$ must represent *all* functions vanishing at the constraints. Equation $f(x) = A(x) + B(x)$ is actually an alternative general form to write a *constrained expression*, that is, an alternative way to generalize interpolation: rather than derive a class of functions (e.g., monomials) satisfying a set of constraints, it represents *all* possible functions satisfying the set of constraints.

To prove that this additive formalism can describe *all* possible functions satisfying the constraints is immediate. Let $f(x)$ be all functions satisfying the constraints and $y(x) = A(x) + B(x)$ be the sum of a specific function satisfying the constraints, $A(x)$, and a function, $B(x)$, representing all functions that are null at the constraints. Then, $y(x)$ will be equal to $f(x)$ iff $B(x) = f(x) - A(x)$, representing all functions that are null at the constraints.

As shown in Equation (5), once the $A(x)$ function is obtained, then the $B(x)$ function can be immediately derived. In fact, $B(x)$ can be obtained by subtracting the $A(x)$ function, where all the constraints are specified in terms of the $g(x)$ free function, from the free function $g(x)$. For this reason, let us write the general expression of a constrained expression as,

$$f(x) = A(x) + g(x) - A(g(x)), \qquad (6)$$

where $A(g(x))$ indicates the function satisfying the constraints where the constraints are specified in term of $g(x)$.

The previous discussion serves to prove that the problem of extending Theory of Connections to higher dimensional spaces consists of the problem of finding the function, $A(x)$, only. In two dimensions, the function $A(x)$ is provided in literature by the Coons surface [11], $f(x,y)$. This surface satisfies the Dirichlet boundary constraints,

$$f(0,y) = c(0,y), \quad f(1,y) = c(1,y), \quad f(x,0) = c(x,0), \quad \text{and} \quad f(x,1) = c(x,1), \tag{7}$$

where the surface is contained in the $x, y \in [0,1] \times [0,1]$ domain. This surface is used in computer graphics and in computational mechanics applications to smoothly join other surfaces together, particularly in finite element method and boundary element method, to mesh problem domains into elements. The expression of the Coons surface is,

$$f(x,y) = (1-x)c(0,y) + x c(1,y) + (1-y) c(x,0) + y c(x,1) - x y c(1,1)$$
$$- (1-x)(1-y) c(0,0) - (1-x) y c(0,1) - x (1-y) c(1,0),$$

where the four subtracting terms are there for continuity. Note the constraint functions at boundary corners must have the same value, $c(0,0)$, $c(0,1)$, $c(1,0)$, and $c(1,1)$. This equation can be written in matrix form as,

$$f(x,y) = \begin{Bmatrix} 1, & 1-x, & x \end{Bmatrix} \begin{bmatrix} 0 & c(x,0) & c(x,1) \\ c(0,y) & -c(0,0) & -c(0,1) \\ c(1,y) & -c(1,0) & -c(1,1) \end{bmatrix} \begin{Bmatrix} 1 \\ 1-y \\ y \end{Bmatrix},$$

or, equivalently,

$$f(x,y) = v^T(x) \, \mathcal{M}(c(x,y)) \, v(y), \tag{8}$$

where

$$\mathcal{M}(c(x,y)) = \begin{bmatrix} 0 & c(x,0) & c(x,1) \\ c(0,y) & -c(0,0) & -c(0,1) \\ c(1,y) & -c(1,0) & -c(1,1) \end{bmatrix} \quad \text{and} \quad v(z) = \begin{Bmatrix} 1 \\ 1-z \\ z \end{Bmatrix}.$$

Since the $f(x,y)$ boundaries match the boundaries of the $c(x,y)$ constraint function, then the identity, $f(x,y) = v^T(x) \, \mathcal{M}(f(x,y)) \, v(y)$, holds for *any* $f(x,y)$ function. Therefore, the $B(x)$ function can be set as,

$$B(x) := g(x,y) - v^T(x) \, \mathcal{M}(g(x,y)) \, v(y), \tag{9}$$

representing all functions that are always zero at the boundary constraints, as $g(x,y)$ is a free function.

4. Theory of Connections Surface Subject to Dirichlet Constraints

Equations (8) and (9) can be merged to provide *all surfaces* with the boundary constraints defined in Equation (7) in the following compact form,

$$f(x,y) = \underbrace{v^T(x)\mathcal{M}(c(x,y))v(y)}_{A(x,y)} + \underbrace{g(x,y) - v^T(x)\mathcal{M}(g(x,y))v(y)}_{B(x,y)}. \tag{10}$$

where, again, $A(x,y)$ indicates an expression satisfying the boundary function constraints defined by $c(x,y)$ and $B(x,y)$ an expression that is zero at the boundaries. In matrix form, Equation (10) becomes,

$$f(x,y) = \begin{Bmatrix} 1 \\ 1-x \\ x \end{Bmatrix}^T \begin{bmatrix} g(x,y) & c(x,0) - g(x,0) & c(x,1) - g(x,1) \\ c(0,y) - g(0,y) & g(0,0) - c(0,0) & g(0,1) - c(0,1) \\ c(1,y) - g(1,y) & g(1,0) - c(1,0) & g(1,1) - c(1,1) \end{bmatrix} \begin{Bmatrix} 1 \\ 1-y \\ y \end{Bmatrix},$$

where $g(x,y)$ is a freely chosen function. In particular, if $g(x,y) = 0$, then the ToC surface becomes the Coons surface.

Figure 5 (left) shows the Coons surface subject to the constraints,

$$c(x,0) = \sin(3x - \pi/4)\,\cos(\pi/3)$$
$$c(x,1) = \sin(3x - \pi/4)\,\cos(4 + \pi/3)$$
$$c(0,y) = \sin(-\pi/4)\,\cos(4y + \pi/3)$$
$$c(1,y) = \sin(3 - \pi/4)\,\cos(4y + \pi/3),$$

and Figure 5 (right) shows a ToC surface that is obtained using the free function,

$$g(x,y) = \frac{1}{3}\cos(4\pi x)\,\sin(6\pi y) - x^2\,\cos(2\pi y). \tag{11}$$

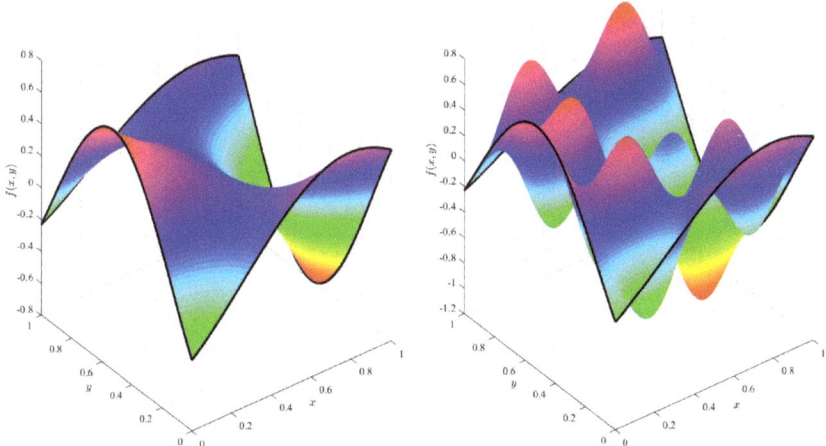

Figure 5. Coons surface (**left**); and ToC surface (**right**) using $g(x,y)$ provided in Equation (11).

For generic boundaries defined in the rectangle $x,y \in [x_i, x_f] \times [y_i, y_f]$, the ToC surface becomes,

$$\begin{aligned}
f(x,y) = g(x,y) &+ \frac{x - x_f}{x_i - x_f}\left[c(x_i, y) - g(x_i, y)\right] + \frac{x - x_i}{x_f - x_i}\left[c(x_f, y) - g(x_f, y)\right] \\
&+ \frac{y - y_f}{y_i - y_f}\left[c(x, y_i) - g(x, y_i)\right] + \frac{y - y_i}{y_f - y_i}\left[c(x, y_f) - g(x, y_f)\right] \\
&- \frac{(x - x_f)(y - y_f)}{(x_i - x_f)(y_i - y_f)}\left[c(x_i, y_i) - g(x_i, y_i)\right] \\
&- \frac{(x - x_f)(y - y_i)}{(x_i - x_f)(y_f - y_i)}\left[c(x_i, y_f) - g(x_i, y_f)\right] \\
&- \frac{(x - x_i)(y - y_f)}{(x_f - x_i)(y_i - y_f)}\left[c(x_f, y_i) - g(x_f, y_i)\right] \\
&- \frac{(x - x_i)(y - y_i)}{(x_f - x_i)(y_f - y_i)}\left[c(x_f, y_f) - g(x_f, y_f)\right].
\end{aligned} \tag{12}$$

Equation (12) can also be set in matrix form,

$$f(x,y) = v_x^\top(x, x_i, x_f)\,\mathcal{M}(x,y)\,v_y(y, y_i, y_f)$$

where
$$M(x,y) = \begin{bmatrix} g(x,y) & c(x,y_i) - g(x,y_i) & c(x,y_f) - g(x,y_f) \\ c(x_i,y) - g(x_i,y) & g(x_i,y_i) - c(x_i,y_i) & g(x_i,y_f) - c(x_i,y_f) \\ c(x_f,y) - g(x_f,y) & g(x_f,y_i) - c(x_f,y_i) & g(x_f,y_f) - c(x_f,y_f) \end{bmatrix}$$

and

$$v_x(x, x_i, x_f) = \begin{Bmatrix} 1 \\ \dfrac{x - x_f}{x_i - x_f} \\ \dfrac{x - x_i}{x_f - x_i} \end{Bmatrix} \quad \text{and} \quad v_y(y, y_i, y_f) = \begin{Bmatrix} 1 \\ \dfrac{y - y_f}{y_i - y_f} \\ \dfrac{y - y_i}{y_f - y_i} \end{Bmatrix}.$$

Note that all the ToC surfaces provided are linear in $g(x,y)$, and, therefore, they can be used to solve, by linear/nonlinear least-squares, two-dimensional optimization problems subject to boundary function constraints, such as linear/nonlinear partial differential equations.

5. Multi-Function Constraints at Generic Coordinates

Equation (12) can be generalized to many function constraints (grid of functions). Assume a set of n_x function constraints $c(x_k, y)$ and a set of n_y function constraints $c(x, y_k)$ intersecting at the $n_x n_y$ points $p_{ij} = c(x_i, y_j)$, then all surfaces satisfying the $n_x n_y$ function constraints can be expressed by,

$$\begin{aligned} f(x,y) = g(x,y) &+ \sum_{k=1}^{n_x} [c(x_k, y) - g(x_k, y)] \prod_{i \neq k} \frac{x - x_i}{x_k - x_i} \\ &+ \sum_{k=1}^{n_y} [c(x, y_k) - g(x, y_k)] \prod_{i \neq k} \frac{y - y_i}{y_k - y_i} \\ &- \sum_{i=1}^{n_x} \left\{ \sum_{j=1}^{n_y} \frac{(x - x_j)(y - y_i)}{(x_i - x_j)(y_j - y_i)} \left[c(x_i, y_j) - g(x_i, y_j) \right] \right\}. \end{aligned} \tag{13}$$

Again, Equation (13) can be written in compact form,

$$f(x,y) = v^\mathsf{T}(x)\, \mathcal{M}(c(x,y))\, v(y) + g(x,y) - v^\mathsf{T}(x)\, \mathcal{M}(g(x,y))\, v(y)$$

where,

$$v(x) = \begin{Bmatrix} 1 \\ \prod_{i \neq 1} \dfrac{x - x_i}{x_1 - x_i} \\ \vdots \\ \prod_{i \neq n_x} \dfrac{x - x_i}{x_{n_x} - x_i} \end{Bmatrix} \quad \text{and} \quad v(y) = \begin{Bmatrix} 1 \\ \prod_{i \neq 1} \dfrac{y - y_i}{y_1 - y_i} \\ \vdots \\ \prod_{i \neq n_y} \dfrac{y - y_i}{y_{n_y} - y_i} \end{Bmatrix}$$

and

$$\mathcal{M}(c(x,y)) = \begin{bmatrix} 0 & c(x, y_1) & \cdots & c(x, y_{n_y}) \\ c(x_1, y) & -c(x_1, y_1) & \cdots & -c(x_1, y_{N_y}) \\ \vdots & \vdots & \ddots & \vdots \\ c(x_{n_x}, y) & -c(x_{n_x}, y_1) & \cdots & -c(x_{n_x}, y_{n_y}) \end{bmatrix}$$

For example, two function constraints in x and three function constraints in y can be obtained using the matrix,

$$\mathcal{M}(c(x,y)) = \begin{bmatrix} 0 & c(x, y_1) & c(x, y_2) & c(x, y_3) \\ c(x_1, y) & -c(x_1, y_1) & -c(x_1, y_2) & -c(x_1, y_3) \\ c(x_2, y) & -c(x_2, y_1) & -c(x_2, y_2) & -c(x_2, y_3) \end{bmatrix}$$

and the vectors,

$$v(x) = \left\{ \begin{array}{c} 1 \\ \dfrac{x-x_2}{x_1-x_2} \\ \dfrac{x-x_1}{x_2-x_1} \end{array} \right\} \quad \text{and} \quad v(y) = \left\{ \begin{array}{c} 1 \\ \dfrac{(y-y_2)(y-y_3)}{(y_1-y_2)(y_1-y_3)} \\ \dfrac{(y-y_1)(y-y_3)}{(y_2-y_1)(y_2-y_3)} \\ \dfrac{(y-y_2)(y-y_1)}{(y_3-y_2)(y_3-y_1)} \end{array} \right\}.$$

Two examples of ToC surfaces are given in Figure 6 in the $x, y \in [-2, 1] \times [1, 3]$ domain.

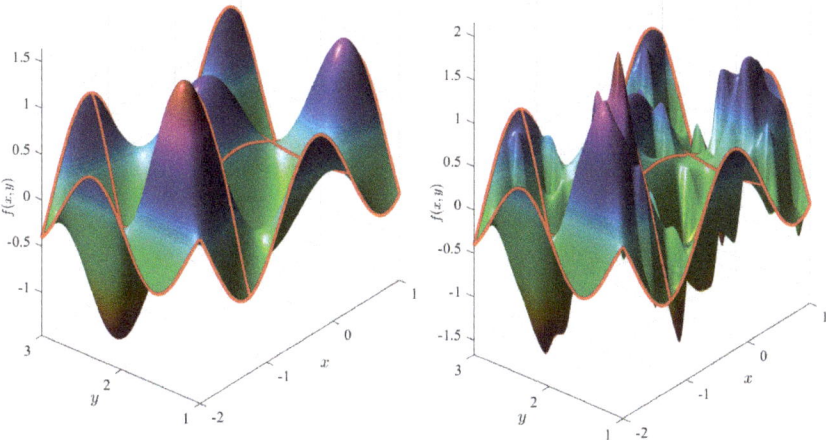

Figure 6. ToC surface subject to multiple constraints on two axes: using $g(x,y) = 0$ (**left**); and using $g(x,y) = \mod(x, 0.5) \cos(19y) - x \mod(3y, 0.4)$ (**right**).

6. Constraints on Function and Derivatives

The "Boolean sum formulation" was provided by Farin [12] (also called "Hermite–Coons formulation") of the Coons surface that includes boundary derivatives,

$$f(x,y) = v^\top(y) F^x(x) + v^\top(x) F^y(y) - v^\top(x) M^{xy} v(y) \qquad (14)$$

where

$$v(z) := \{2z^3 - 3z^2 + 1, \quad z^3 - 2z^2 + z, \quad -2z^3 + 3z^2, \quad z^3 - z^2\}^\top$$
$$F^x(x) := \{c(x,0), \quad c_y(x,0), \quad c(x,1), \quad c_y(x,1)\}^\top$$
$$F^y(y) := \{c(0,y), \quad c_x(0,y), \quad c(1,y), \quad c_x(1,y)\}^\top$$

and

$$M^{xy}(x,y) := \begin{bmatrix} c(0,0) & c_y(0,0) & c(0,1) & c_y(0,1) \\ c_x(0,0) & c_{xy}(0,0) & c_x(0,1) & c_{xy}(0,1) \\ c(1,0) & c_y(1,0) & c(1,1) & c_y(1,1) \\ c_x(1,0) & c_{xy}(1,0) & c_x(1,1) & c_{xy}(1,1) \end{bmatrix}.$$

The formulation provided in Equation (14) can be put in the matrix compact form,

$$f(x,y) = v^\top(x) \, \mathcal{M}(c(x,y)) \, v(y), \qquad (15)$$

where
$$v(z) := \{1,\quad 2z^3 - 3z^2 + 1,\quad z^3 - 2z^2 + z,\quad -2z^3 + 3z^2,\quad z^3 - z^2\}^\top \tag{16}$$

and the 5×5 matrix, $\mathcal{M}(c(x,y))$, has the expression,

$$\mathcal{M}(c(x,y)) := \begin{bmatrix} 0 & c(x,0) & c_y(x,0) & c(x,1) & c_y(x,1) \\ c(0,y) & -c(0,0) & -c_y(0,0) & -c(0,1) & -c_y(0,1) \\ c_x(0,y) & -c_x(0,0) & -c_{xy}(0,0) & -c_x(0,1) & -c_{xy}(0,1) \\ c(1,y) & -c(1,0) & -c_y(1,0) & -c(1,1) & -c_y(1,1) \\ c_x(1,y) & -c_x(1,0) & -c_{xy}(1,0) & -c_x(1,1) & -c_{xy}(1,1) \end{bmatrix}. \tag{17}$$

To verify the boundary derivative constraints, the following partial derivatives of Equation (15) are used,

$$f_x(x,y) = [v_x^\top(x)\mathcal{M}(c(x,y)) + v^\top(x)\mathcal{M}_x(c(x,y))]v(y)$$
$$f_y(x,y) = v^\top(x)[\mathcal{M}_y^\top(c(x,y))v(y) + \mathcal{M}(c(x,y))v_y(y)],$$

where

$$\frac{dv}{dz} = \begin{Bmatrix} 0 \\ 6z(z-1) \\ 3z^2 - 4z + 1 \\ 6z(1-z) \\ z(3z-2) \end{Bmatrix}, \quad \mathcal{M}_y = \begin{bmatrix} 0 & 0_{1\times 4} \\ c_y(0,y) & 0_{1\times 4} \\ c_{xy}(0,y) & 0_{1\times 4} \\ c_y(1,y) & 0_{1\times 4} \\ c_{xy}(1,y) & 0_{1\times 4} \end{bmatrix}, \quad \text{and} \quad \mathcal{M}_x^\top = \begin{bmatrix} 0 & 0_{1\times 4} \\ c_x(x,0) & 0_{1\times 4} \\ c_{xy}(x,0) & 0_{1\times 4} \\ c_x(x,1) & 0_{1\times 4} \\ c_{xy}(x,1) & 0_{1\times 4} \end{bmatrix}.$$

The ToC in 2D with function and derivative boundary constraints is simply,

$$f(x,y) = \underbrace{v^\top(x)\mathcal{M}(c(x,y))v(y)}_{A(x,y)} + \underbrace{g(x,y) - v^\top(x)\mathcal{M}(g(x,y))v(y)}_{B(x,y)} \tag{18}$$

where the $\mathcal{M}$ matrix and the v vectors are provided by Equations (17) and (16), respectively.

Dirichlet/Neumann mixed constraints can be derived, as shown in the examples provided in Sections 6.1–6.4. The matrix compact form is simply obtained from the matrix defined in Equation (17) by removing the rows and the columns associated with the boundary constraints not provided, while the vectors $v(x)$ and $v(y)$ are derived by specifying the constraints. Note that in general the vectors $v(x)$ and $v(y)$ are *not* unique. The reason the vectors $v(x)$ and $v(y)$ are not unique comes from the fact that the $A(x)$ term in Equation (6) is not unique.

In the next subsections, four Dirichlet/Neumann mixed constraint examples providing the simplest expressions for $v(x)$ and $v(y)$ are derived. The Appendix A contains the expressions for the $v(x)$ and $v(y)$ vectors associated with all the combinations of Dirichlet and Neumann constraints.

6.1. Constraints: $c(0,y)$ and $c(x,0)$

In this case, the Coons-type surface satisfying the boundary constraints can be expressed as,

$$f(x,y) = \begin{Bmatrix} 1 & p(x) \end{Bmatrix} \begin{bmatrix} 0 & c(x,0) \\ c(0,y) & -c(0,0) \end{bmatrix} \begin{Bmatrix} 1 \\ q(y) \end{Bmatrix}$$

where $p(x)$ and $q(y)$ are unknown functions. Expanding, we obtain $f(x,y) = c(x,0)q(y) + p(x)[c(0,y) - c(0,0)q(y)]$. The two constraints are satisfied if,

$$c(0,y) = c(0,0)q(y) + p(0)[c(0,y) - c(0,0)q(y)]$$
$$c(x,0) = c(x,0)q(0) + p(x)[c(0,0) - c(0,0)q(0)].$$

Therefore, the $p(x)$ and $q(y)$ functions must satisfy $p(0) = 1$ and $q(0) = 1$. The simplest expressions satisfying these equations can be obtained by selecting $p(x) = 1$ and $q(y) = 1$. In this case, the associated ToC surface is given by,

$$f(x,y) = \{1 \quad 1\} \begin{bmatrix} g(x,y) & c(x,0) - g(x,0) \\ c(0,y) - g(0,y) & g(0,0) - c(0,0) \end{bmatrix} \begin{Bmatrix} 1 \\ 1 \end{Bmatrix}$$

Note that any functions satisfying $p(0) = 1$ and $q(0) = 1$ can be adopted to obtain the ToC surface satisfying the constraints $f(0,y) = c(0,y)$ and $f(x,0) = c(x,0)$. This is because there are infinite Coons-type surfaces satisfying the constraints. Consequently, the vectors $v(x)$ and $v(y)$ are not unique.

6.2. Constraints: $c(0,y)$ and $c_y(x,0)$

For these boundary constraints, the Coons-type surface is expressed by,

$$f(x,y) = \{1 \quad p(x)\} \begin{bmatrix} 0 & c_y(x,0) \\ c(0,y) & -c_y(0,0) \end{bmatrix} \begin{Bmatrix} 1 \\ q(y) \end{Bmatrix}$$
$$= c_y(x,0)q(y) + p(x)[c(0,y) - c_y(0,0)q(y)].$$

The constraints are satisfied if,

$$c(0,y) = c_y(0,0)q(y) + p(0)[c(0,y) - c_y(0,0)q(y)],$$
$$c_y(x,0) = c_y(x,0)q_y(0) + p(x)[c_y(0,0) - c_y(0,0)q_y(0)].$$

Therefore, the $p(x)$ and $q(y)$ functions must satisfy $p(0) = 1$ and $q_y(0) = 1$. One solution is $p(x) = 1$ and $q(y) = y$. Therefore, the associated ToC surface is given by,

$$f(x,y) = \{1 \quad 1\} \begin{bmatrix} g(x,y) & c_y(x,0) - g_y(x,0) \\ c(0,y) - g(0,y) & g_y(0,0) - c_y(0,0) \end{bmatrix} \begin{Bmatrix} 1 \\ y \end{Bmatrix}.$$

6.3. Neumann Constraints: $c_x(0,y)$, $c_x(1,y)$, $c_y(x,0)$, and $c_y(x,1)$

In this case, the Coons-type surface satisfying the boundary constraints can be expressed as,

$$f(x,y) = \{1, \quad p_1(x), \quad p_2(x)\} \begin{bmatrix} 0 & c_y(x,0) & c_y(x,1) \\ c_x(0,y) & -c_{xy}(0,0) & -c_{xy}(0,1) \\ c_x(1,y) & -c_{xy}(1,0) & -c_{xy}(1,1) \end{bmatrix} \begin{Bmatrix} 1 \\ q_1(y) \\ q_2(y) \end{Bmatrix}.$$

The constraints are satisfied if,

$$c_x(0,y) = q_1(y)c_{xy}(0,0) + q_2(y)c_{xy}(0,1) +$$
$$+ p_{1x}(0)[c_x(0,y) - q_1(y)c_{xy}(0,0) - q_2(y)c_{xy}(0,1)] +$$
$$+ p_{2x}(0)[c_x(1,y) - q_1(y)c_{xy}(1,0) - q_2(y)c_{xy}(1,1)]$$
$$c_x(1,y) = q_1(y)c_{xy}(1,0) + q_2(y)c_{xy}(1,1) +$$
$$+ p_{1x}(1)[c_x(0,y) - q_1(y)c_{xy}(0,0) - q_2(y)c_{xy}(0,1)] +$$
$$+ p_{2x}(1)[c_x(1,y) - q_1(y)c_{xy}(1,0) - q_2(y)c_{xy}(1,1)]$$
$$c_y(x,0) = q_{1y}(0)c_y(x,0) + q_{2y}(0)c_y(x,1) +$$
$$+ p_1(x)[c_{xy}(0,0) - q_{1y}(0)c_{xy}(0,0) - q_{2y}(0)c_{xy}(0,1)] +$$
$$+ p_2(x)[c_{xy}(1,0) - q_{1y}(0)c_{xy}(1,0) - q_{2y}(0)c_{xy}(1,1)]$$
$$c_y(x,1) = q_{1y}(1)c_y(x,0) + q_{2y}(1)c_y(x,1) +$$
$$+ p_1(x)[c_{xy}(0,1) - q_{1y}(1)c_{xy}(0,0) - q_{2y}(1)c_{xy}(0,1)] +$$
$$+ p_2(x)[c_{xy}(1,1) - q_{1y}(1)c_{xy}(1,0) - q_{2y}(1)c_{xy}(1,1)].$$

These equations imply $p_{1x}(0) = q_{1x}(0) = 1$, $p_{1x}(1) = q_{1x}(1) = 0$, $p_{2x}(0) = q_{2x}(0) = 0$, and $p_{2x}(1) = q_{2x}(1) = 1$. Therefore, the simplest solution is $p_1(t) = q_1(t) = t - t^2/2$ and $p_2(t) = q_2(t) = t^2/2$. Then, the associated ToC surface satisfying the Neumann constraints is given by,

$$f(x,y) = v^\mathsf{T}(x) \begin{bmatrix} g(x,y) & c_y(x,0) - g_y(x,0) & c_y(x,1) - g_y(x,1) \\ c_x(0,y) - g_x(0,y) & g_{xy}(0,0) - c_{xy}(0,0) & g_{xy}(0,1) - c_{xy}(0,1) \\ c_x(1,y) - g_x(1,y) & g_{xy}(1,0) - c_{xy}(1,0) & g_{xy}(1,1) - c_{xy}(1,1) \end{bmatrix} v(y)$$

where

$$v^\mathsf{T}(x) = \left\{1,\ x - \frac{x^2}{2},\ \frac{x^2}{2}\right\} \quad \text{and} \quad v(y) = \left\{1,\ y - \frac{y^2}{2},\ \frac{y^2}{2}\right\}.$$

6.4. Constraints: $c(0,y)$, $c_y(x,0)$, and $c_y(x,1)$

In this case, the Coons-type surface satisfying the boundary constraints is in the form,

$$f(x,y) = \begin{Bmatrix} 1 \\ p(x) \end{Bmatrix}^\mathsf{T} \begin{bmatrix} 0 & c_y(x,0) & c_y(x,1) \\ c(0,y) & -c_y(0,0) & -c_y(0,1) \end{bmatrix} \begin{Bmatrix} 1 \\ q_1(y) \\ q_2(y) \end{Bmatrix}.$$

The constraints are satisfied if $p(0) = 1$, $p_{1y}(0) = 1$, $p_{1y}(1) = 0$, $p_{2y}(0) = 0$, and $p_{2y}(1) = 1$. Therefore, the associated ToC surface is,

$$f(x,y) = \begin{Bmatrix} 1 \\ 1 \end{Bmatrix}^\mathsf{T} \begin{bmatrix} g(x,y) & c_y(x,0) - g_y(x,0) & c_y(x,1) - g_y(x,1) \\ c(0,y) - g(0,y) & g_y(0,0) - c_y(0,0) & g_y(0,1) - c_y(0,1) \end{bmatrix} \begin{Bmatrix} 1 \\ y - \frac{y^2}{2} \\ \frac{y^2}{2} \end{Bmatrix}.$$

6.5. Generic Mixed Constraints

Consider the case of mixed constraints,

$$\begin{aligned} f(x,y_1) &= c(x,y_1) \\ f_x(x,y_2) &= c_x(x,y_2) \\ f(x,y_3) &= c(x,y_3) \end{aligned} \quad \text{and} \quad \begin{aligned} f_y(x_1,y) &= c_y(x_1,y) \\ f_y(x_2,y) &= c_y(x_2,y) \\ f(x_3,y) &= c(x_3,y) \end{aligned} . \tag{19}$$

In this case, the surface satisfying the boundary constraints is built using the matrix,

$$\mathcal{M}(c(x,y)) = \begin{bmatrix} 0 & c(x,y_1) & c_x(x,y_2) & c(x,y_3) \\ c_y(x_1,y) & -c_y(x_1,y_1) & -c_{xy}(x_1,y_2) & -c_y(x_1,y_3) \\ c_y(x_2,y) & -c_y(x_2,y_1) & -c_{xy}(x_2,y_2) & -c_y(x_2,y_3) \\ c(x_3,y) & -c(x_3,y_1) & -c_x(x_3,y_2) & -c(x_3,y_3) \end{bmatrix}$$

and all surfaces subject to the constraints defined in Equation (19) can be obtained by,

$$f(x,y) = v(x)^\mathsf{T} \mathcal{M}(c(x,y)) v(y) + g(x,y) - v(x)^\mathsf{T} \mathcal{M}(g(x,y)) v(y),$$

where

$$v(x) = \begin{Bmatrix} 1 \\ p_1(x,x_1,x_2,x_3) \\ p_2(x,x_1,x_2,x_3) \\ p_3(x,x_1,x_2,x_3) \end{Bmatrix} \quad \text{and} \quad v(y) = \begin{Bmatrix} 1 \\ q_1(y,y_1,y_2,y_3) \\ q_2(y,y_1,y_2,y_3) \\ q_3(y,y_1,y_2,y_3) \end{Bmatrix}$$

are vectors made of the (not unique) function vectors $v(x)$ and $v(y)$ whose expressions can be found by satisfying the constraints (as done in the previous four subsections) along with a methodology similar to that given in Section 5.

7. Extension to n-Dimensional Spaces and Arbitrary-Order Derivative Constraints

This section provides the *Multivariate Theory of Connections*, as the generalization to n-dimensional rectangular domains with arbitrary-order boundary derivatives of what is presented above for two-dimensional space. Using tensor notation, this generalization is represented in the following compact form,

$$F(x) = \underbrace{\mathcal{M}(c(x))_{i_1 i_2 \ldots i_n} v_{i_1} v_{i_2} \ldots v_{i_n}}_{A(x)} + \underbrace{g(x) - \mathcal{M}(g(x))_{i_1 i_2 \ldots i_n} v_{i_1} v_{i_2} \ldots v_{i_n}}_{B(x)} \tag{20}$$

where n is the number of orthogonal coordinates defined by the vector $x = \{x_1, x_2, \ldots, x_n\}$, $v_{i_k}(x_k)$ is the i_kth element of a vector function of the variable x_k, $\mathcal{M}$ is an n-dimensional tensor that is a function of the boundary constraints defined in $c(x)$, and $g(x)$ is the free-function.

In Equation (20), the term $A(x)$ represents any function satisfying the boundary constraints defined by $c(x)$ and the term $B(x)$ represents all possible functions that are zero on the boundary constraints. The subsections that follow explain how to construct the $\mathcal{M}$ tensor and the v_{i_k} vectors for assigned boundary constraints, and provides a proof that the tensor formulation of the ToC defined by Equation (20) satisfies all boundary constraints defined by $c(x)$, independently of the choice of the free function, $g(x)$.

Consider a generic boundary constraint on the $x_k = p$ hyperplane, where $k \in [1, n]$. This constraint specifies the d-derivative of the constraint function $c(x)$ evaluated at $x_k = p$ and it is indicated by $^k c_p^d := \left. \dfrac{\partial^d c(x)}{\partial x_k^d} \right|_{x_k = p}$. Consider a set of ℓ_k constraints defined in various x_k hyperplanes. This set of constraints is indicated by $^k c_{p^k}^{d^k}$, where d^k and p^k are vectors of ℓ_k elements indicating the order of derivatives and the values of x_k where the boundary constraints are defined, respectively. A specific boundary constraint, e.g. the mth boundary constraint, can then be written as $^k c_{p_m^k}^{d_m^k}$.

Additionally, let us define an operator, called the boundary constraint operator, whose purpose is to take the dth derivative with respect to coordinate x_k and then evaluate that function at $x_k = p$. Equation (21) shows the idea.

$$^k b_p^d [f] \equiv \left. \frac{\partial^d f}{\partial x_k^d} \right|_{(x_1, \ldots, x_{k-1}, p, x_{k+1}, \ldots, x_n)} \tag{21}$$

In general, for a function of n variables, the boundary constraint operator identifies an $n-1$-dimensional manifold. As the boundary constraint operator is used throughout this proof, it is important to note its properties when acting on sums and products of functions. Equation (22) shows how the boundary constraint operator acts on sums, and Equation (23) shows how the boundary constraint operator acts on products.

$$^kb_p^d[f_1+f_2] = {}^kb_p^d[f_1] + {}^kb_p^d[f_2] \tag{22}$$

$$^kb_p^d[f_1 f_2] = \begin{cases} {}^kb_p^d[f_1]\,{}^kb_p^d[f_2], & d=0 \\ {}^kb_p^d[f_1]f_2 + f_1\,{}^kb_p^d[f_2], & d>0 \end{cases} \tag{23}$$

This section shows how to build the $\mathcal{M}$ tensor and the vectors v given the boundary constraints defined by the boundary constraint operators. Moreover, this section contains a proof that, in Equation (20), the boundary constraints defined by $c(x)$ satisfy the function $A(x)$ and, by extension, the function $B(x)$ projects the free-function $g(x)$ onto the sub-space of functions that are zero on the boundary constraints. Then, it follows that the expression for the ToC surface given in Equation (20) represents *all* possible functions that meet the boundary defined by the boundary constraint operators.

7.1. The $\mathcal{M}$ Tensor

There is a step-by-step method for constructing the $\mathcal{M}$ tensor.

1. The element of $\mathcal{M}$ for all indices equal to 1 is 0 (i.e., $\mathcal{M}_{11\ldots1} = 0$).
2. The first order tensor obtained by keeping the kth dimension's index and setting all other dimension's indices to 1 can be written as,

$$\mathcal{M}_{1,\ldots,1,i_k,1,\ldots,1} = {}^kc_{p^k}^{d^k}, \qquad \text{where} \quad i_k \in [2, \ell_k+1],$$

where the vector ${}^kc_{p^k}^{d^k}$ contains the ℓ_k boundary constraints specified along the x_k-axis. For example, consider the following $\ell_7 = 3$ constraints on the $k = 7$th axis,

$$^7c_{p^7}^{d^7} := \left\{ c|_{x_7=-0.3},\ \left.\frac{\partial^4 c}{\partial x_7^4}\right|_{x_7=0.5},\ \left.\frac{\partial c}{\partial x_7}\right|_{x_7=1.1} \right\} \quad \text{then}: \begin{cases} d^7 = \{0,\ 4,\ 1\} \\ p^7 = \{-0.3,\ 0.5,\ 1.1\}. \end{cases}$$

3. The generic element of the tensor is $\mathcal{M}_{i_1 i_2 \ldots i_n}$, where at least two indices are different from 1. Let m be the number of indices different from 1. Note that m is also the number of constraint "intersections". In this case, the generic element of the $\mathcal{M}$ tensor is provided by,

$$\mathcal{M}_{i_1 i_2 \ldots i_n} = {}^1b_{p_{i_1-1}^1}^{d_{i_1-1}^1}\left[{}^2b_{p_{i_2-1}^2}^{d_{i_2-1}^2}\left[\ldots\left[{}^nb_{p_{i_n-1}^n}^{d_{i_n-1}^n}[c(x)]\right]\ldots\right]\right](-1)^{m+1}. \tag{24}$$

If $c(x) \in C^s$, where $s = \sum_{k=1}^{n} d_{i_k-1}^k$, then Clairaut's theorem states that the sequence of boundary constraint operators provided in Equation (24) can be freely permutated. This permutation becomes obvious by multiple applications of the theorem. For example,

$$f_{xyy} = (f_{xy})_y = (f_{yx})_y = (f_y)_{xy} = (f_y)_{yx} = f_{yyx}.$$

To better clarify how to use Equation (24), consider the example of the following constraints in three-dimensional space.

$$c(x)|_{x_1=0},\quad c(x)|_{x_1=1},\quad c(x)|_{x_2=0},\quad \left.\frac{\partial c(x)}{\partial x_2}\right|_{x_2=0},\quad c(x)|_{x_3=0},\quad \text{and}\quad \left.\frac{\partial c(x)}{\partial x_3}\right|_{x_3=0}$$

1. From Step 1: $M_{111} = 0$
2. From Step 2:

$$M_{i_111} = \left\{0,\ c(0, x_2, x_3),\ c(1, x_2, x_3)\right\} = \left\{0,\ {}^1b_0^0[c(x)],\ {}^1b_1^0[c(x)]\right\}$$

$$M_{1i_21} = \left\{0,\ c(x_1, 0, x_3),\ \frac{\partial c}{\partial x_2}(x_1, 0, x_3)\right\} = \left\{0,\ {}^2b_0^0[c(x)],\ {}^2b_0^1[c(x)]\right\}$$

$$M_{11i_3} = \left\{0,\ c(x_1, x_3, 0),\ \frac{\partial c}{\partial x_3}(x_1, x_2, 0)\right\} = \left\{0,\ {}^3b_0^0[c(x)],\ {}^3b_0^1[c(x)]\right\}$$

3. From Step 3, a single example is provided,

$$M_{323} = {}^1b_1^0\left[{}^2b_0^0\left[{}^3c_0^1(x)\right]\right](-1)^4 = \left.\frac{\partial c(x)}{\partial x_3}\right|_{\substack{x_1=1 \\ x_2=0 \\ x_3=0}}$$

which, thanks to Clairaut's theorem, can also be written as,

$$M_{323} = {}^2b_0^0\left[{}^3b_0^1\left[{}^1c_1^0\right]\right](-1)^4 = {}^3b_0^1\left[{}^1b_1^0\left[{}^2c_0^0\right]\right](-1)^4.$$

Three additional examples are given to help further illustrate the procedure,

$$M_{132} = -\left.\frac{\partial c(x)}{\partial x_2}\right|_{\substack{x_2=0 \\ x_3=0}}, \quad M_{221} = -c(0, 0, x_3), \quad \text{and} \quad M_{333} = \left.\frac{\partial^2 c(x)}{\partial x_2 \partial x_3}\right|_{\substack{x_1=1 \\ x_2=0 \\ x_3=0}}$$

*7.2. The **v** Vectors*

Each vector, v_k, is associated with the ℓ_k constraints that are specified by ${}^kc_{p^k}^{d^k}$. The v_k vector is built as follows,

$$v_k = \left\{1,\ \sum_{i=1}^{\ell_k} \alpha_{i1} h_i(x_k),\ \sum_{i=1}^{\ell_k} \alpha_{i2} h_i(x_k),\ \ldots,\ \sum_{i=1}^{\ell_k} \alpha_{i\ell_k} h_i(x_k)\right\},$$

where $h_i(x_k)$ are ℓ_k linearly independent functions. The simplest set of linearly independent functions are monomials, that is, $h_i(x_k) = x_k^{i-1}$. The $\ell_k \times \ell_k$ coefficients, α_{ij}, can be computed by matrix inversion,

$$\begin{bmatrix} {}^kb_{p_1}^{d_1}[h_1] & {}^kb_{p_1}^{d_1}[h_2] & \cdots & {}^kb_{p_1}^{d_1}[h_{\ell_k}] \\ {}^kb_{p_2}^{d_2}[h_1] & {}^kb_{p_2}^{d_2}[h_2] & \cdots & {}^kb_{p_2}^{d_2}[h_{\ell_k}] \\ \vdots & \vdots & \ddots & \vdots \\ {}^kb_{p_{\ell_k}}^{d_{\ell_k}}[h_1] & {}^kb_{p_{\ell_k}}^{d_{\ell_k}}[h_2] & \cdots & {}^kb_{p_{\ell_k}}^{d_{\ell_k}}[h_{\ell_k}] \end{bmatrix} \begin{bmatrix} \alpha_{11} & \alpha_{12} & \cdots & \alpha_{1\ell_k} \\ \alpha_{21} & \alpha_{22} & \cdots & \alpha_{2\ell_k} \\ \vdots & \vdots & \ddots & \vdots \\ \alpha_{\ell_k 1} & \alpha_{\ell_k 2} & \cdots & \alpha_{\ell_k \ell_k} \end{bmatrix} = \begin{bmatrix} 1 & 0 & \cdots & 0 \\ 0 & 1 & \cdots & 0 \\ \vdots & \vdots & \ddots & \vdots \\ 0 & 0 & \cdots & 1 \end{bmatrix}. \quad (25)$$

To supplement the above explanation, let us look at the example of Dirichlet boundary conditions on x_1 from the example in Section 7.1. There are two boundary conditions, $c(x)|_{x_1=0}$ and $c(x)|_{x_1=1}$, and thus two linearly independent functions are needed,

$$v_{i_1} = \left\{1,\ \alpha_{11} h_1(x_1) + \alpha_{21} h_2(x_1),\ \alpha_{12} h_1(x_1) + \alpha_{22} h_2(x_1)\right\}.$$

Let us consider, $h_1(x_1) = 1$ and $h_2(x_1) = x_1$. Then, following Equation (25),

$$\begin{bmatrix} {}^1b_0^0[1] & {}^1b_0^0[x] \\ {}^2b_1^0[1] & {}^2b_1^0[x] \end{bmatrix} \begin{bmatrix} \alpha_{11} & \alpha_{12} \\ \alpha_{21} & \alpha_{22} \end{bmatrix} = \begin{bmatrix} 1 & 0 \\ 1 & 1 \end{bmatrix} \begin{bmatrix} \alpha_{11} & \alpha_{12} \\ \alpha_{21} & \alpha_{22} \end{bmatrix} = \begin{bmatrix} 1 & 0 \\ 0 & 1 \end{bmatrix} \rightarrow \begin{bmatrix} \alpha_{11} & \alpha_{12} \\ \alpha_{21} & \alpha_{22} \end{bmatrix} = \begin{bmatrix} 1 & 0 \\ -1 & 1 \end{bmatrix},$$

and substituting the values of α_{ij}, we obtain $v_{i_1} = \left\{1, \ 1-x_1, \ x_1\right\}$.

7.3. Proof

This section demonstrates that the term $A(x)$ from Equation (20) generates a surface satisfying the boundary constraints defined by the function $c(x)$. First, it is shown that $A(x)$ satisfies boundary constraints on the value, and then that $A(x)$ satisfies boundary constraints on arbitrary-order derivatives.

Equation (23) for $d=0$ allows us to write,

$$^k b^0_{p_{q-1}}[A(x)] = {}^k b^0_{p_{q-1}}[\mathcal{M}_{i_1 i_2 \ldots i_k \ldots i_n}] v_{i_1} v_{i_2} \ldots {}^k b^0_{p_{q-1}}[v_{i_k}] \ldots v_{i_n}. \tag{26}$$

The boundary constraint operator applied to v_k yields,

$$^k b^0_{p_{q-1}}[v_{i_k}] = \begin{cases} = 1, & i_k = 1, q \\ = 0, & i_k \neq 1, q. \end{cases} \tag{27}$$

Since the only nonzero terms are associated with $i_k = 1, q$, we have,

$$^k b^0_{p_{q-1}}[A(x)] = \left({}^k b^0_{p_{q-1}}[\mathcal{M}_{i_1 i_2 \ldots 1 \ldots i_n}] + {}^k b^0_{p_{q-1}}[\mathcal{M}_{i_1 i_2 \ldots q \ldots i_n}] \right) v_{i_1} v_{i_2} \ldots v_{i_n}. \tag{28}$$

Applying the boundary constraint operator to the $n-1$-dimensional $\mathcal{M}$ tensor where index $i_k = q$ has no effect, because all of the functions already have coordinate x_k substituted for the value p_{q-1} (see Equation (24)). Moreover, applying the boundary constraint operator to the $\mathcal{M}$ tensor where index $i_k = 1$ causes all terms in the sum within the parenthesis in Equation (28) to cancel each other, except when all of the non-i_k indices are equal to one. This leads to Equation (29).

$$^k b^0_{p_{q-1}}[A(x)] = \left(\mathcal{M}_{11\ldots 1\ldots 1} + \mathcal{M}_{11\ldots q\ldots 1} \right) v_1 v_1 \ldots v_1 \tag{29}$$

Since $v_j = 1$ when $j=1$ and $\mathcal{M}_{11\ldots 1} = 0$ by definition, then,

$$^k b^0_{p_{q-1}}[A(x)] = \mathcal{M}_{11\ldots q\ldots 1} = c(x_1, x_2, \ldots, p_{q-1}, \ldots, x_n),$$

which proves Equation (20) works for boundary constraints on the value.

Now, we show that Equation (20) holds for arbitrary-order derivative type boundary constraints. Equation (23) for $d>0$ allows us to write,

$$^k b^{d_{q-1}}_{p_{q-1}}[A(x)] = {}^k b^{d_{q-1}}_{p_{q-1}}[\mathcal{M}_{i_1 i_2 \ldots i_k \ldots i_n}] v_{i_1} v_{i_2} \ldots v_{i_k} \ldots v_{i_n} + \mathcal{M}_{i_1 i_2 \ldots i_k \ldots i_n} v_{i_1} v_{i_2} \ldots {}^k b^{d_{q-1}}_{p_{q-1}}[v_{i_k}] \ldots v_{i_n}. \tag{30}$$

From Equation (23), we note that boundary constraint operators that take a derivative follow the usual product rule when applied to a product. Moreover, we note that all of the v vectors except v_{i_k} do not depend on x_k, thus applying the boundary constraint operator to them results in a vector of zeros. Applying the boundary constraint operator to v_{i_k} yields,

$$^k b^{d_{q-1}}_{p_{q-1}}[v_{i_k}] = \begin{cases} = 1, & i_k = q \\ = 0, & i_k \neq q, \end{cases}$$

and applying the boundary constraint operator to $\mathcal{M}$ yields,

$$k b_{p_{q-1}}^{d_{q-1}}[\mathcal{M}_{i_1 i_2 \ldots 1 \ldots i_n}] = \begin{cases} = k b_{p_{q-1}}^{d_{q-1}}[\mathcal{M}_{i_1 i_2 \ldots 1 \ldots i_n}], & i_k = 1 \\ = 0, & i_k \neq 1. \end{cases}$$

Substituting these simplifications into $A(x) = \mathcal{M}_{i_1 i_2 \ldots i_k \ldots i_n} v_{i_1} v_{i_2} \ldots v_{i_k} \ldots v_{i_n}$, after applying the boundary constraint operator, results in Equation (31).

$$k b_{p_{q-1}}^{d_{q-1}}[A(x)] = \left(k b_{p_{q-1}}^{d_{q-1}}[\mathcal{M}_{i_1 i_2 \ldots 1 \ldots i_n}] + \mathcal{M}_{i_1 i_2 \ldots q \ldots i_n} \right) v_{i_1} v_{i_2} \ldots v_{i_n} \quad (31)$$

Similar to the proof for value-based boundary constraints, based on Equation (24), all terms in the sum within the parenthesis in Equation (31) cancel each other, except when all of the non-i_k indices are equal to one. Thus, Equation (31) can be simplified to Equation (32).

$$k b_{p_{q-1}}^{d_{q-1}}[A(x)] = \left(k b_{p_{q-1}}^{d_{q-1}}[\mathcal{M}_{11 \ldots 1 \ldots 1}] + \mathcal{M}_{11 \ldots q \ldots 1} \right) v_1 v_1 \ldots v_1 \quad (32)$$

Again, all of the vectors v were designed such that their first component is 1, and the value of the element of $\mathcal{M}$ for all indices equal to 1 is 0. Therefore, Equation (32) simplifies to,

$$k b_{p_{q-1}}^{d_{q-1}}[A(x)] = \mathcal{M}_{11 \ldots q \ldots 1} = \left. \frac{\partial^d c(x)}{\partial x_k^d} \right|_{x_k = p_{q-1}},$$

which proves Equation (20) works for arbitrary-order derivative boundary constraints.

In conclusion, the term $A(x)$ from Equation (20) generates a manifold satisfying the boundary constraints given in terms of arbitrary-order derivative in n-dimensional space. The term $B(x)$ from Equation (20) projects any free function $g(x)$ onto the space of functions that are vanishing at the specified boundary constraints. As a result, Equation (20) can be used to produce the family of *all* possible functions satisfying assigned boundary constraints (functions or derivatives) in rectangular domains in n-dimensional space.

8. Conclusions

This paper extends to n-dimensional spaces the Univariate Theory of Connections (ToC), introduced in Ref. [1]. First, it provides a mathematical tool to express *all* possible surfaces subject to constraint functions and arbitrary-order derivatives in a boundary rectangular domain, and then it extends the results to the multivariate case by providing the Multivariate Theory of Connections, which allows one to obtain n-dimensional manifolds subject to any-order derivative boundary constraints.

In particular, if the constraints are provided along one axis only, then this paper shows that the univariate ToC, as defined in Ref. [1], can be adopted to describe *all* possible surfaces satisfying the constraints. If the boundary constraints are defined in a rectangular domain, then the constrained expression is found in the form $f(x) = A(x) + B(x)$, where $A(x)$ can be *any* function satisfying the constraints and $B(x)$ describes *all* functions that are vanishing at the constraints. This is obtained by introducing a free function, $g(x)$, into the function $B(x)$ in such a way that $B(x)$ is zero at the constraints no matter what the $g(x)$ is. This way, by spanning all possible $g(x)$ surfaces (even discontinuous, null, or piece-wise defined) the resulting $B(x)$ generates *all* surfaces that are zero at the constraints and, consequently, $f(x) = A(x) + B(x)$, describes all surfaces satisfying the constraints defined in the rectangular boundary domain. The function $A(x)$ has been selected as a Coons surface [11] and, in particular, a Coons surface is obtained if $g(x) = 0$ is selected. All possible combinations of Dirichlet *and* Neumann constraints are also provided in Appendix A.

The last section provides the Multivariate Theory of Connections extension, which is a mathematical tool to transform n-dimensional constraint optimization problems subject to constraints on the boundary value and any-order derivative into unconstrained optimization problems. The number of applications of the Multivariate Theory of Connections are many, especially in the area of partial and stochastic differential equations: the main subjects of our current research.

Author Contributions: C.L. derived the table in Appendix A and the mathematical proof validating the tensor notation. All the remaining parts are provided by D.M.

Funding: This research received no external funding.

Acknowledgments: The authors acknowledge Ergun Akleman for pointing out the Coons surface.

Conflicts of Interest: The authors declare no conflict of interest.

Abbreviations

The following abbreviation is used in this manuscript:

ToC   Theory of Connections
PDE   Partial Differential Equations
ODE   Ordinary Differential Equations
IVP   Initial Value Problems
BVP   Boundary Value Problems

Appendix A. All combinations of Dirichlet and Neumann constraints

$c_{x,0}$	$c_{0,y}$	$c_{x,1}$	$c_{1,y}$	$c^x_{0,y}$	$c^x_{1,y}$	$c^y_{x,0}$	$c^y_{x,1}$	$v(x)$	$v(y)$
✓	✓							$\{1, 1\}$	$\{1, 1\}$
	✓					✓		$\{1, 1\}$	$\{1, y\}$
			✓			✓		$\{1, x\}$	$\{1, y\}$
✓	✓	✓						$\{1, 1\}$	$\{1, 1-y^2, y^2\}$
✓	✓						✓	$\{1, 1\}$	$\{1, 1, y\}$
✓			✓	✓				$\{1, x\}$	$\{1, 1-y, y\}$
			✓	✓		✓		$\{1, x\}$	$\{1, y-y^2, y^2\}$
		✓				✓	✓	$\{1, 1\}$	$\{1, y-y^2/2, y^2/2\}$

$c_{x,0}$	$c_{0,y}$	$c_{x,1}$	$c_{1,y}$	$c^x_{0,y}$	$c^x_{1,y}$	$c^y_{x,0}$	$c^y_{x,1}$	$v(x)$	$v(y)$
					✓	✓	✓	$\begin{Bmatrix}1\\x\end{Bmatrix}$	$\begin{Bmatrix}1\\y-y^2/2\\y^2/2\end{Bmatrix}$
✓	✓	✓	✓					$\begin{Bmatrix}1\\1-x\\x\end{Bmatrix}$	$\begin{Bmatrix}1\\1-y\\y\end{Bmatrix}$
	✓	✓	✓			✓		$\begin{Bmatrix}1\\1-x\\x\end{Bmatrix}$	$\begin{Bmatrix}1\\y-y^2\\y^2\end{Bmatrix}$
	✓		✓			✓	✓	$\begin{Bmatrix}1\\1-x\\x\end{Bmatrix}$	$\begin{Bmatrix}1\\y-y^2/2\\y^2/2\end{Bmatrix}$
		✓	✓	✓		✓		$\begin{Bmatrix}1\\x-x^2\\x^2\end{Bmatrix}$	$\begin{Bmatrix}1\\y-y^2\\y^2\end{Bmatrix}$
		✓		✓		✓	✓	$\begin{Bmatrix}1\\x-x^2\\x^2\end{Bmatrix}$	$\begin{Bmatrix}1\\y-y^2/2\\y^2/2\end{Bmatrix}$
			✓	✓	✓	✓	✓	$\begin{Bmatrix}1\\x-x^2/2\\x^2/2\end{Bmatrix}$	$\begin{Bmatrix}1\\y-y^2/2\\y^2/2\end{Bmatrix}$
✓	✓					✓		$\begin{Bmatrix}1\\1\end{Bmatrix}$	$\begin{Bmatrix}1\\1\\y\end{Bmatrix}$
✓						✓		$\begin{Bmatrix}1\\x\end{Bmatrix}$	$\begin{Bmatrix}1\\1\\y\end{Bmatrix}$
✓	✓		✓			✓		$\begin{Bmatrix}1\\1-x\\x\end{Bmatrix}$	$\begin{Bmatrix}1\\1\\y\end{Bmatrix}$
✓	✓	✓				✓		$\begin{Bmatrix}1\\1\end{Bmatrix}$	$\begin{Bmatrix}1\\1-y^2\\y-y^2\\y^2\end{Bmatrix}$
✓		✓	✓			✓		$\begin{Bmatrix}1\\x\end{Bmatrix}$	$\begin{Bmatrix}1\\1-y^2\\y-y^2\\y^2\end{Bmatrix}$
✓	✓					✓	✓	$\begin{Bmatrix}1\\1\end{Bmatrix}$	$\begin{Bmatrix}1\\1\\y-y^2/2\\y^2/2\end{Bmatrix}$

$c_{x,0}$	$c_{0,y}$	$c_{x,1}$	$c_{1,y}$	$c^x_{0,y}$	$c^x_{1,y}$	$c^y_{x,0}$	$c^y_{x,1}$	$v(x)$	$v(y)$
✓					✓	✓	✓	$\left\{\begin{array}{c}1\\x\end{array}\right\}$	$\left\{\begin{array}{c}1\\1\\y-y^2/2\\y^2/2\end{array}\right\}$
✓	✓					✓	✓	$\left\{\begin{array}{c}1\\1\\x\end{array}\right\}$	$\left\{\begin{array}{c}1\\1\\y\end{array}\right\}$
✓				✓	✓	✓		$\left\{\begin{array}{c}1\\x-x^2/2\\x^2/2\end{array}\right\}$	$\left\{\begin{array}{c}1\\1\\y\end{array}\right\}$
✓	✓			✓		✓		$\left\{\begin{array}{c}1\\1\\x\end{array}\right\}$	$\left\{\begin{array}{c}1\\1\\y\end{array}\right\}$
✓	✓	✓		✓		✓		$\left\{\begin{array}{c}1\\1\\x\end{array}\right\}$	$\left\{\begin{array}{c}1\\1-y^2\\y-y^2\\y^2\end{array}\right\}$
✓	✓				✓	✓	✓	$\left\{\begin{array}{c}1\\1\\x\end{array}\right\}$	$\left\{\begin{array}{c}1\\1\\y-y^2/2\\y^2/2\end{array}\right\}$
✓	✓	✓	✓			✓		$\left\{\begin{array}{c}1\\1-x\\x\end{array}\right\}$	$\left\{\begin{array}{c}1\\1-y^2\\y-y^2\\y^2\end{array}\right\}$
✓		✓	✓	✓			✓	$\left\{\begin{array}{c}1\\x-x^2\\x^2\end{array}\right\}$	$\left\{\begin{array}{c}1\\1-y^2\\y-y^2\\y^2\end{array}\right\}$
✓	✓		✓			✓	✓	$\left\{\begin{array}{c}1\\1-x\\x\end{array}\right\}$	$\left\{\begin{array}{c}1\\1\\y-y^2/2\\y^2/2\end{array}\right\}$
✓			✓	✓		✓	✓	$\left\{\begin{array}{c}1\\x-x^2\\x^2\end{array}\right\}$	$\left\{\begin{array}{c}1\\1\\y-y^2/2\\y^2/2\end{array}\right\}$
✓				✓	✓	✓	✓	$\left\{\begin{array}{c}1\\x-x^2/2\\x^2/2\end{array}\right\}$	$\left\{\begin{array}{c}1\\1\\y-y^2/2\\y^2/2\end{array}\right\}$

$c_{x,0}$	$c_{0,y}$	$c_{x,1}$	$c_{1,y}$	$c^x_{0,y}$	$c^x_{1,y}$	$c^y_{x,0}$	$c^y_{x,1}$	$v(x)$	$v(y)$
✓	✓	✓	✓	✓		✓		$\begin{Bmatrix} 1 \\ 1-x^2 \\ x-x^2 \\ x^2 \end{Bmatrix}$	$\begin{Bmatrix} 1 \\ 1-y^2 \\ y-y^2 \\ y^2 \end{Bmatrix}$
✓	✓		✓	✓		✓	✓	$\begin{Bmatrix} 1 \\ 1-x^2 \\ x-x^2 \\ x^2 \end{Bmatrix}$	$\begin{Bmatrix} 1 \\ 1 \\ y-y^2/2 \\ y^2/2 \end{Bmatrix}$
✓	✓			✓	✓	✓	✓	$\begin{Bmatrix} 1 \\ 1 \\ x-x^2/2 \\ x^2/2 \end{Bmatrix}$	$\begin{Bmatrix} 1 \\ 1 \\ y-y^2/2 \\ y^2/2 \end{Bmatrix}$
✓	✓	✓	✓			✓	✓	$\begin{Bmatrix} 1 \\ 1-x \\ x \end{Bmatrix}$	$\begin{Bmatrix} 1 \\ 1-3y^2+2y^3 \\ y-2y^2+y^3 \\ 3y^2-2y^3 \\ -y^2+y^3 \end{Bmatrix}$
✓		✓	✓	✓		✓	✓	$\begin{Bmatrix} 1 \\ -1+x \\ 1 \end{Bmatrix}$	$\begin{Bmatrix} 1 \\ 1-3y^2+2y^3 \\ y-2y^2+y^3 \\ 3y^2-2y^3 \\ -y^2+y^3 \end{Bmatrix}$
✓			✓	✓	✓	✓	✓	$\begin{Bmatrix} 1 \\ x-x^2/2 \\ x^2/2 \end{Bmatrix}$	$\begin{Bmatrix} 1 \\ 1-3y^2+2y^3 \\ y-2y^2+y^3 \\ 3y^2-2y^3 \\ -y^2+y^3 \end{Bmatrix}$
✓	✓	✓		✓		✓	✓	$\begin{Bmatrix} 1 \\ 1 \\ x \end{Bmatrix}$	$\begin{Bmatrix} 1 \\ 1-3y^2+2y^3 \\ y-2y^2+y^3 \\ 3y^2-2y^3 \\ -y^2+y^3 \end{Bmatrix}$
✓	✓	✓	✓	✓		✓	✓	$\begin{Bmatrix} 1 \\ 1-x^2 \\ x-x^2 \\ x^2 \end{Bmatrix}$	$\begin{Bmatrix} 1 \\ 1-3y^2+2y^3 \\ y-2y^2+y^3 \\ 3y^2-2y^3 \\ -y^2+y^3 \end{Bmatrix}$
✓	✓	✓		✓	✓	✓	✓	$\begin{Bmatrix} 1 \\ 1 \\ x-x^2/2 \\ x^2/2 \end{Bmatrix}$	$\begin{Bmatrix} 1 \\ 1-3y^2+2y^3 \\ y-2y^2+y^3 \\ 3y^2-2y^3 \\ -y^2+y^3 \end{Bmatrix}$

$c_{x,0}$	$c_{0,y}$	$c_{x,1}$	$c_{1,y}$	$c_{0,y}^x$	$c_{1,y}^x$	$c_{x,0}^y$	$c_{x,1}^y$	$v(x)$	$v(y)$
✓	✓	✓	✓	✓	✓	✓	✓	$\begin{Bmatrix} 1 \\ 1 - 3x^2 + 2x^3 \\ x - 2x^2 + x^3 \\ 3x^2 - 2x^3 \\ -x^2 + x^3 \end{Bmatrix}$	$\begin{Bmatrix} 1 \\ 1 - 3y^2 + 2y^3 \\ y - 2y^2 + y^3 \\ 3y^2 - 2y^3 \\ -y^2 + y^3 \end{Bmatrix}$

References

1. Mortari, D. The theory of connections: Connecting points. *Mathematics* **2017**, *5*, 57. [CrossRef]
2. Johnston, H.; Mortari, D. Linear differential equations subject to relative, integral, and infinite constraints. In Proceedings of the Astrodynamics Specialist Conference, Snowbird, UT, USA, 19–23 August 2018; pp. 18–273.
3. Johnston, H.; Mortari, D. Weighted least-squares solutions of over-constrained differential equations. In Proceedings of the IAA SciTech-081 Forum on Space Flight Mechanics and Space Structures and Materials, Moscow, Russia, 13–15 November 2018.
4. Mortari, D. Least-squares solutions of linear differential equations. *Mathematics* **2017**, *5*, 48. [CrossRef]
5. Mortari, D.; Johnston, H.; Smith, L. High accurate least-squares solutions of nonlinear differential equations. *J. Comput. Appl. Math.* **2018**, *352*, 293–307. [CrossRef]
6. *MATLAB and Statistics Toolbox Release 2012b*; The MathWorks, Inc.: Natick, MA, USA, 2012.
7. *Chebfun Guide*; Driscoll, T.A., Hale, N., Trefethen, L.N., Eds.; Chebfun Guide Pafnuty Publications: Oxford, UK, 2014.
8. Mortari, D.; Furfaro, R. Theory of connections applied to first-order system of ordinary differential equations subject to component constraints. In Proceedings of the 2018 AAS/AIAA Astrodynamics Specialist Conference, Snowbird, UT, USA, 19–23 August 2018.
9. Furfaro, R.; Mortari, D. Least-squares solution of a class of optimal guidance problems. In Proceedings of the 2018 AAS/AIAA Astrodynamics Specialist Conference, Snowbird, UT, USA, 19–23 August 2018.
10. Waring, E. Problems concerning interpolations. *Philos. Trans. R. Soc. Lond.* **1779**, *69*, 59–67.
11. Coons, S.A. *Surfaces for Computer Aided Design*; Technical Report; Massachusetts Institute of Technology: Cambridge, MA, USA, 1964.
12. Farin, G. *Curves and Surfaces for CAGD: A Practical Guide*, 5th ed.; Morgan Kaufmann Publishers Inc.: San Francisco, CA, USA, 2002.

© 2019 by the authors. Licensee MDPI, Basel, Switzerland. This article is an open access article distributed under the terms and conditions of the Creative Commons Attribution (CC BY) license (http://creativecommons.org/licenses/by/4.0/).

Article

Learning Algorithms for Coarsening Uncertainty Space and Applications to Multiscale Simulations

Zecheng Zhang [1], Eric T. Chung [2], Yalchin Efendiev [1,3,*] and Wing Tat Leung [4]

[1] Department of Mathematics, Texas A&M University, College Station, TX 77843, USA; tom_z_z@tamu.edu
[2] Department of Mathematics, The Chinese University of Hong Kong, Shatin, New Territories, Hong Kong, China; tschung@math.cuhk.edu.hk
[3] Multiscale Model Reduction Laboratory, North-Eastern Federal University, Yakutsk 677980, Russia
[4] Institute for Computational Engineering and Sciences, The University of Texas at Austin, Austin, TX 78712, USA; sidnet123@gmail.com
* Correspondence: efendiev@math.tamu.edu

Received: 21 March 2020; Accepted: 24 April 2020; Published: 4 May 2020

Abstract: In this paper, we investigate and design multiscale simulations for stochastic multiscale PDEs. As for the space, we consider a coarse grid and a known multiscale method, the generalized multiscale finite element method (GMsFEM). In order to obtain a small dimensional representation of the solution in each coarse block, the uncertainty space needs to be partitioned (coarsened). This coarsening collects realizations that provide similar multiscale features as outlined in GMsFEM (or other method of choice). This step is known to be computationally demanding as it requires many local solves and clustering based on them. In this work, we take a different approach and learn coarsening the uncertainty space. Our methods use deep learning techniques in identifying clusters (coarsening) in the uncertainty space. We use convolutional neural networks combined with some techniques in adversary neural networks. We define appropriate loss functions in the proposed neural networks, where the loss function is composed of several parts that includes terms related to clusters and reconstruction of basis functions. We present numerical results for channelized permeability fields in the examples of flows in porous media.

Keywords: generalized multiscale finite element method; multiscale model reduction; clustering; deep learning

1. Introduction

Many problems are multiscale with uncertainties. Examples include problems in porous media, material sciences, biological sciences, and so on. For example, in porous media applications, engineers can obtain fine-scale data about pore geometries or subsurface properties at very fine resolutions. These data are obtained in some spatial locations and then generalized to the entire reservoir domain. As a result, one uses geostatistical or other statistical tools to populate the media properties in space. The resulting porous media properties are stochastic and one needs to deal with many porous media realizations, where each realization is multiscale and varies at very fine scales. There are other realistic problems which have multiscale properties with uncertainties such as the multiscale public safety systems, [1], multiscale social networks [2]; these problems usually have more data.

Simulating each realization can be computationally expensive because of the media's multiscale nature. Our objective is to simulate many of these realizations. To address the issues associated with spatial and temporal scales, many multiscale methods have been developed [3–12]. These methods perform simulations on the coarse grid by developing reduced-order models. However, developing reduced-order models requires local computations, which can be expensive when one deals with many realizations. For this reason, some type of coarsening of the uncertainty space is needed [13]. In this paper, we consider some novel approaches for developing coarsening of uncertainty space as discussed below.

To coarsen the uncertainty space, clustering algorithms are often used; but a proper distance function should be designed in order to make the clusters have physical sense and achieve a reduction in the uncertainty space. The paper [13] proposed a method that uses the distance between local solutions. The motivation is that the local problems with random boundary conditions can represent the main models with all boundary conditions. Due to a high dimension of the uncertainty space, the authors in [13] proposed to compute the local solutions of only several realizations and then use the Karhunen–Loeve expansion [14] to approximate the solutions of all the other realizations. The distance function is then defined to be the distance between solutions and the standard K-means [15] algorithm is used to cluster the uncertainty space.

The issue with this method is computing the local solutions in the local neighborhoods. It is computationally expensive to compute the local solutions; although the KL expansion can save time to approximate the solutions of other realizations, one still needs to decide how many selected realizations we need to represent all the other solutions. In this paper, we propose the use of deep learning methodology and avoid explicit clustering as in earlier works. We remark that the development of deep learning techniques for multiscale simulations are recently reported in [16–20].

In this work, to coarsen the uncertainty space, we propose a deep learning algorithm which will learn the clusters for each local neighborhood. Due the nature of the permeability fields, we can use the transfer learning which uses the parameters of one local neighborhood to initialize the learning of all the other neighborhoods. This saves significantly computational time.

The auto encoder structure [21] has been widely used in improving the K-mean clustering algorithm [22–24]. The idea is to use the encoder to extract features and reduce the dimension; the encoding process can also be taken as a kernel method [25] which maps the data to a space which is easier to be separated. The decoder is used to upsample the latent space (reduced dimension feature space) back to the input space. The clustering algorithm is then used to cluster the latent space, which will save time due to the low dimension of the latent space and also preserve the accuracy due to the features extracted by the encoder.

Traditionally, the learning process is only involved in reconstructing the input space. Such kind of methods ignore the features extracted by latent space; so, it is not clear if the latent space is good enough to represent the input space and is easily clustered by the K-means method. In [24], the authors proposed a new loss which includes the reconstruction loss meanwhile the loss results from the clustering. The authors claimed that the new loss improves the clustering results.

We will apply the auto encoder structure and the multiple loss function; however, we will design the auto encoder as a generative network, i.e., the input and output space are different. More precisely, the input is the uncertain space (permeability fields) and the output will be the multiscale functions co-responding to the uncertain space. Intuitively, we want to use the multiscale basis to supervise the learning of the clusters so that the clusters will inherit the property of the solution. The motivation is the multiscale basis can somehow represent the real solutions and permeability fields; hence, the latent space is no longer good for clustering the input space but will be suitable for representing the multiscale basis function space.

To define the reconstructing loss, the common idea is the mean square error (MSE); but many works [26–29] have shown that the MSE tends to produce the average effect. In fact, in the area of image super-resolution [26–36] and other low level computer vision tasks, the generated images are usually over-smooth if trained using MSE. The theory is the MSE will capture the low frequency features like the background which is relatively steady; but for images with high contrast, the MSE will usually try to blur the images and the resulting images will lose the colorfulness and become less vivid [26]. Our problem has multiscale nature and we want to capture the dominant modes and multiscale features, hence a single MSE is clearly not enough.

Following the idea from [27,29], we consider adding an adversary net [37]. The motivation is the fact that different layers of fully convolutional network extract different features [29,38,39]. Deep fully convolutional neural networks (FCN) [40–45] have demonstrated its power in almost all computer vision tasks. Convolution operation is a local operation and the network with full convolutions are independent with the input size. People now are clear about the functioning of the different layers of the FCN. In computer vision task, the lower layers (layers near input) tend to generate sharing features of all objects like edges and curves while the higher layers (near output) are more object oriented. If we train the network using the loss from the lower layers, the texture and details are persevered, while the higher layers will keep the general spatial structure.

This motivates us using the losses from different layers of the fully convolutional layers. Multiple layers will give us a multilevel capture of the basis features and hence measure the basis in a more complete way. To implement the idea, we will pretrain an adversary net; and input the multiscale basis of the generative net and the real basis. The losses then come from some selected layers of the adversary net. Although it is still not clear the speciality of each layer, if we consider the multiscale physical problem, the experiments show that the accuracy is improved and, amazingly, the training becomes easier when compared to the MSE of the basis directly.

The uncertain space coarsening (cluster) is performed using the deep learning idea described above. Due to the space dimension, we will perform the clustering algorithm locally in space; that is, we first need a spatial coarsening. Due to the multiscale natural of the problem, this motivates us using the generalized multiscale finite element methods (GMsFEM) which derive the multiscale basis of a coarse neighborhood by solving the local problem. GMeFEM was first proposed in [46] and further studied in [3–10]. This method is a generalization of the multiscale finite element method [47,48]. The work starts from constructing the snapshot space for each local neighborhood. The snapshot space is constructed by solving local problems and several methods including harmonic extension, random boundary condition [49] have been proposed. Once we have the snapshot space, the offline space which will be used as computing the solution are constructed by using spectral decomposition.

Our method is designed for solving PDEs with heterogeneous properties and uncertainty. The heterogeneity and uncertainty in our models come from the permeability $\kappa(x,s)$. To verify our method, we numerically simulate around 240,000 local spatial fields which contain complex information such as the moving channels. Our model is then trained and tested based on the generated spatial fields. It should be noted that our method could be applied to some other realistic problems which contain large-scale data such as detecting extreme values with order statistics in samples from continuous distributions [50], as well as to some other subjects, e.g., multiscale social networks [2] and the multiscale public safety systems [1]. These topic will be studied in the future.

The rest of the work is organized as follow: in Section 2, we consider the problem setup and introduce both uncertain space and spatial coarsening. In Section 3, we introduce the structure of the network and the training algorithm. In Section 4, we will present the numerical results. The paper ends with conclusions.

2. Problem Settings

In this section, we will present some basic ideas involving the use of the generalized multiscale finite element method (GMsFEM) for parameter-dependent problems. Let D be a bounded domain in $\mathbb{R}^2$ and Ω be the parameter space in $\mathbb{R}^N$. We consider the following parameter-dependent elliptic problem:

$$-\nabla \cdot (\kappa(x,s)\nabla u(x,s)) = f(x,s), (x,s) \in D \times \Omega, \tag{1}$$

$$u(x,s) = 0, (x,s) \in \partial D \times \Omega, \tag{2}$$

where $\kappa(x,s)$ is a heterogeneous coefficient depending on both the spatial variable x and the parameter s, and $f \in L^2(D)$ is a given source. We remark that the differential operators in Equation (1) are defined with respect to the spatial variable x. This is the case for the rest of the paper.

2.1. The Coarsening of the Parameter Space. The Main Idea

The parameter space Ω is assumed to be of very high dimension (i.e., large N) and consists of very large number of realizations. For a given realization, the idea is to find its representation in the coarse space and use the coarse space to perform the computation. We will use the deep cluster learning algorithm to perform the coarsening. Due to the heterogeneous properties of the proposed problem, fine mesh is used; this will bring difficulties in coarsening the parameter space and in computation of the solution. We hence perform the parameter coarsening locally in the space D, i.e., we also coarsen the spatial domain. To coarsen the spatial domain, we use coarse grids and consider the GMsFEM.

In Figure 1, we present an illustration of the proposed coarsening technique. On the left figure, the coarse grid blocks in the space are shown. Each coarse grid has a different cluster in the uncertainty space Ω, which corresponds to the coarsening of the uncertainty space. The main objective in multiscale methods is efficiently finding the clustering of the uncertainty space, which is our main goal.

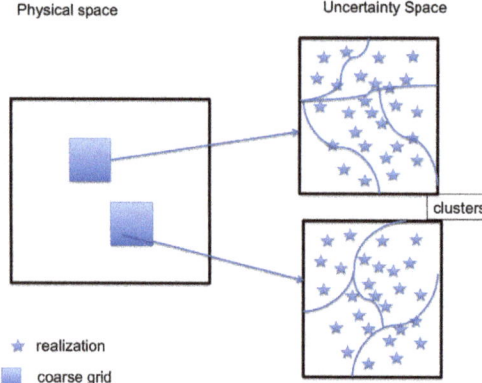

Figure 1. Illustration of coarsening of space and uncertainties. Different clusters for different coarse blocks. On the left plot, two coarse blocks are shown. On the right plot, clusters are illustrated.

2.2. Space Coarsening—Generalized Multiscale Finite Element Method

It is computationally expensive to capture heterogeneous properties using very fine mesh. For this reason, we use GMsFEM to coarsen the spatial representation of the solution. The coarsening of the parameter space will be performed in each local spatial neighborhood. We will achieve this goal by the GMsFEM, which will briefly be discussed. Consider the second order elliptic equation $Lu = f$ in D with proper boundary conditions; denote the the elliptic operator as:

$$L(u) = -\frac{\partial}{\partial x_i}(k_{ij}(x)\frac{\partial}{\partial x_j}u). \qquad (3)$$

Let the spatial domain D be partitioned by a coarse grid $\mathcal{T}^H$; this does not resolve the multiscale features. Let us denote K as one cell in $\mathcal{T}^H$ and refine K to obtain the fine grid partition $\mathcal{T}^h$ (blue box in Figure 2). We assume the fine grid is a conforming refinement of the coarse grid. See Figure 2 for details.

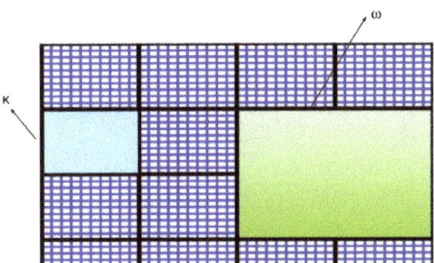

Figure 2. Domain Partition $\mathcal{T}^H$.

For the i-th coarse grid node, let ω_i be the set of all coarse elements having the vertex i (green region in Figure 2). We will solve local problem in each coarse neighborhood to obtain set of multiscale basis functions $\{\phi_i^{\omega_i}\}$ and seek solution in the form:

$$u = \sum_i \sum_j c_{ij} \phi_j^{\omega_i}, \qquad (4)$$

where $\phi_j^{\omega_i}$ is the offline basis function in the i-th coarse neighborhood ω_i and j denotes the j-th basis function. Before we construct the offline basis, we first need to derive the snapshot basis.

2.2.1. Snapshot Space

There are several methods to construct the snapshot space; we will use the harmonic extension of the fine grid functions defined on the boundary of ω_i. Let us denote $\delta_l^h(x)$ as fine grid delta function, which is defined as $\delta_l^h(x_k) = \delta_{l,k}$ for $x_k \in J_h(\omega_i)$ where $J_h(\omega_i)$ denotes the boundary nodes of ω_i. The snapshot function $\psi_l^{\omega_i}$ is then calculated by solving local problem in ω_i:

$$L(\psi_l^{\omega_i}) = 0 \qquad (5)$$

subject to the boundary condition $\psi_l^{\omega_i} = \delta_l^h(x)$. The snapshot space $V_{snap}^{\omega_i}$ is then constructed as the span of all snapshot functions.

2.2.2. Offline Spaces

The offline space $V_{off}^{\omega_i}$ is derived from the snapshot space and is used for computing the solution of the problem. We need to solve for a spetral problem and this can be summarized as finding λ and $v \in V_{snap}^{\omega_i}$ such that:

$$a_{\omega_i}(v,w) = \lambda s_{\omega_i}(v,w), \forall w \in V_{snap}^{\omega_i}, \qquad (6)$$

where a_{ω_i} is symmetric non-negative definite bilinear form and s_{ω_i} is symmetric positive definite bilinear form. By convergence analysis, they are given by

$$a_{\omega_i}(v,w) = \int_{\omega_i} \kappa \nabla v \cdot \nabla w, \qquad (7)$$

$$s_{\omega_i}(v,w) = \int_{\omega_i} \tilde{\kappa} v \cdot w. \qquad (8)$$

In the above definition of s_{ω_i}, the function $\tilde{\kappa} = \kappa \sum |\nabla \chi_j|^2$ where $\{\chi_j\}$ is a set of partition of unity functions corresponding to the coarse grid partition of the domain D and the summation is taken over all the functions in this set. The offline space is then constructed by choosing the smallest L_i eigenvalues and we can form the space by the linear combination of snapshot basis using corresponding eigenvectors:

$$\phi_k^{\omega_i} = \sum_{j=1}^{L_i} \Psi_{kj}^{\omega_i} \psi_j^{\omega_i}, \qquad (9)$$

where $\Psi_{kj}^{\omega_i}$ is the jth element of kth eigenvector and L_i is the number of snapshot basis. V_{off} is then defined as the collection of all local offline basis functions. Finally we are trying to find $u_{off} \in V_{off}$ such that

$$a(u_{off}, v) = \int_D fv, \forall v \in V_{off} \qquad (10)$$

where $a(u,v) = \int_D \kappa \nabla u \cdot \nabla v$. For more details, we refer the readers to the references [8–10].

2.3. The Idea of the Proposed Method

We present the general methodology in this section. The target is to save the time in computing the GMsFEM basis $\phi_k^{\omega_i}$ for all ω_i and for all uncertain space parameters. We propose the clustering algorithm to coarsen the uncertain space in each local neighborhood. The key to the success of the clustering is that: the cluster should inherit the property of the solution, that is, the local heterogeneous fields $\kappa(x,s)$ clustered into the same group should have similar solution properties. When the cluster is learned by the some learning algorithm, the only computation involved is to fit the local neighborhood of the given testing heterogeneous field into some cluster. This is a feed forward process including several convolution operations and matrix multiplications and compared to the direct computing, we save a lot of time in computing the spectral problem in Equation (6) and the inverse of a matrix Equation (10). The detailed process is illustrated in the following chart (Figure 3):

1. (Training) For a given input local neighborhood ω_j, we train the cluster (which will be detailed in next section) of the parameter space Ω and get the clusters $S_1^j, ..., S_n^j$, where n is the number of clusters and is uniform for all j. Please note that we may have different cluster assignments in different local neighborhoods.
2. (Training) For each local neighborhood ω_j and cluster S_i^j, define the average $\bar{\kappa}_{ij}$ and compute generalized multiscale basis for $\bar{\kappa}_{ij}$.
3. (Testing) Given a new $\kappa(x,s)$ and for each local neighborhood ω_j, fit $\kappa(x,s)$ into a S_i^j by the trained network (step 1) and use the pre-computed GMsFEM basis (step 2) to find the solution.

It should be noted that we perform clustering using the heterogeneous fields; however, the cluster should inherit the property of the solution corresponding to the heterogeneous fields. This makes the clustering challenging. The performance of the standard K-means algorithm relies on the initialization and the distance metric. We may initialize the algorithm based on the clustering of the heterogeneous fields but we need to design a good metric. In the next section, we are going to introduce a learning algorithm which uses an auto-encoder structure and multiple losses to achieve the required clustering task.

Mathematics **2020**, *8*, 720

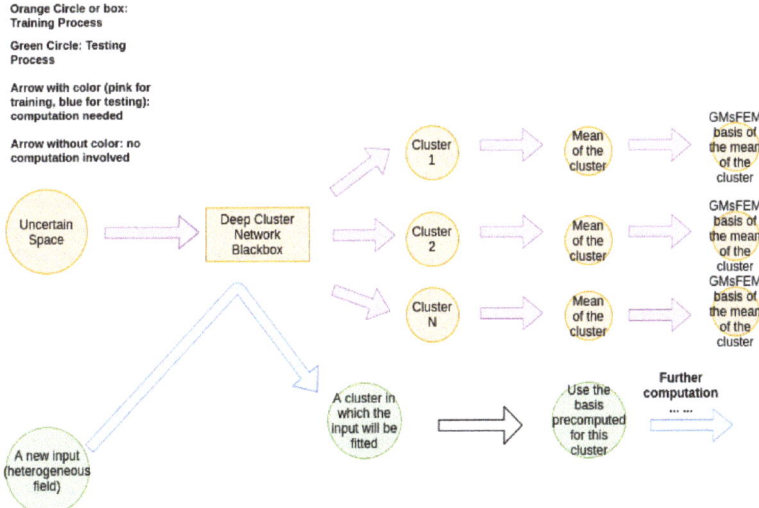

Figure 3. Work flow of the proposed method.

3. Deep Learning

The network is consisted of two sub networks. The first one is targeted to performing the clustering and the second one, which is the adversary net, will serve as the reconstruction of loss function.

3.1. Clustering Net

The cluster net is aimed for clustering the heterogeneous fields $\kappa(x,s)$; but the resulting clusters should inherit the properties of the solution corresponding to the $\kappa(x,s)$, i.e., the heterogeneous fields grouped in the same cluster should have similar corresponding solution properties. This similarity will be measured by the adversary net which will be introduced in Section 3.3. We hence design the network demonstrated in Figure 4.

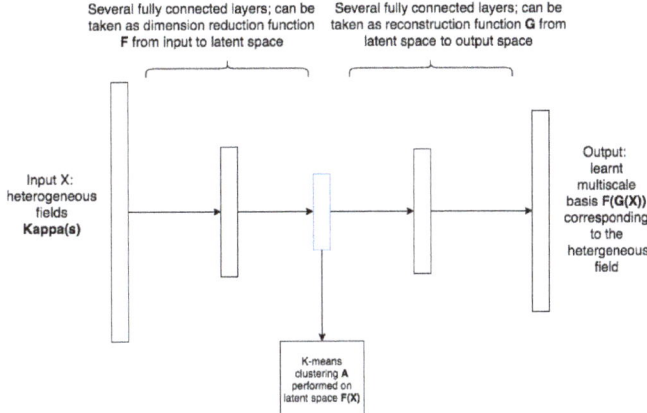

Figure 4. Cluster network.

The input $X \in \mathbb{R}^{m,d}$, where m is the number of samples and d is the dimension of one local heterogeneous field, of the network is the local heterogeneous fields which are parametrized by the random variable $s \in \Omega$. The output of the network is the multiscale basis (first GMsFEM basis) which somehow represents the solution corresponding to the coefficient $\kappa(x,s)$. This is a generative network which has an auto encoder structure. The dimension reduction function $F(X)$ can be interpreted as some kind of kernel method which maps the input data to a new space which is easier to be separated. This can also be interpreted as the learning of a good metric function for the later performed K-mean clustering. We will perform K-means clustering algorithm in latent space $F(X)$. $G(\cdot)$ will then transfer the latent space data to the space of multiscale basis function. This process can be taken as a generative process and we reconstruct the basis from the extracted features. The detailed algorithm is as follow (see Figure 5 for an illustration):

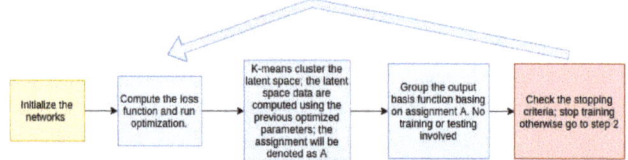

Figure 5. Deep learning algorithm.

Steps illustrated in Figure 5:

1. Initialize the networks and clustering the output basis function.
2. Compute the loss function L (defined later) and run optimization.
3. Cluster the latent space by K-means algorithm (reduced dimension space, which is a middle layer of the cluster network); the latent space data are computed using the previous optimized parameters; the assignment will be denoted as A.
4. Basis functions whose corresponding inputs are in the same cluster (basing on assignment A) will be grouped together. No training or fitting-in involved in this step.
5. Repeat step 2 to step 4 until the stopping criteria is met.

3.2. Loss Functions

Loss function is the key to the deep learning. Our loss function is consisted of cluster loss and the reconstruction loss.

1. Clustering loss $C(\theta_F, \theta_G)$: this is the mean standard deviation of all clusters of the learned basis and θ is the parameters we need to optimize. It should be noted that the loss here is computed using the learned basis instead of the input of the network. This loss controls the clustering process, i.e., the smaller the loss, the better the clustering in the sense of clustering the multiscale basis. Let us denote κ_{ij} as jth realization in ith cluster; $G(F(\kappa_{ij})) \in \mathbb{R}^d$ will then be jth learned basis in cluster i and let θ_G and θ_F be the parameters associated with G and F, the loss is then defined as follow,

$$C(\theta_F, \theta_G) = \frac{1}{|A|} \sum_i^{|A|} \sum_j^{A_i} \frac{1}{A_i} \|G(F(\kappa_{ij})) - \bar{\phi}_i\|_2^2, \quad (11)$$

where $|A|$ is the number of clusters which is a hyper parameter and A_i denotes the number of elements in cluster i; $\bar{\phi}_i \in \mathbb{R}^d$ is the mean of cluster i. This loss clearly serves the purpose of clustering the solution instead of the input heterogeneous fields; however, in order to guarantee the learned basis are closed to the pre-computed multiscale basis, we need to define the reconstruction loss which measures the difference between the learned basis and the pre-computed basis.

2. Reconstruction loss $R(\theta_F, \theta_G)$: this is the mean square error of multiscale basis $Y \in \mathbb{R}^{m,d}$, where m is the number of samples. This loss controls the construction process, i.e., if the loss is small,

the learned basis are close to the real multiscale basis. This loss will supervise the learning of the cluster. It is defined as follow:

$$R(\theta_F, \theta_G) = \frac{1}{m} \sum_i^m \|G(F(\kappa_i)) - \phi_i\|_2^2, \tag{12}$$

where $G(F(\kappa_i)) \in \mathbb{R}^d$ and $\phi_i \in \mathbb{R}^d$ are learned and pre-computed multiscale basis of ith sample κ_i separately.

The entire loss function is then defined as $L(\theta_F, \theta_G) = \lambda_1 C + \lambda_2 R$, where λ_1, λ_2 are predefined weights. We are going to solve the following optimization problem:

$$\min_{\theta_G, \theta_F} L(\theta_F, \theta_G) \tag{13}$$

for the required training process.

3.3. Adversary Network Severing as an Additional Loss

We have introduced the reconstruction loss which measures the similarity between the learned basis and the pre-computed basis in the previous section. It is the mean square error (MSE) of the learned and pre-computed basis. MSE is a smooth loss and easy to train but there is a well known fact about MSE that this loss will blur the image. In the area of image super-resolution and other low level computer vision tasks, the loss is not friendly to inputs with high contrast and the resulting generated images are usually over smooth. Our problem has multiscale nature and is similar with the low level computer vision task, i.e., this is a generative task; hence blurring and over smoothing should happen if the model is trained by MSE. To define a great reconstruction loss is important.

Motivated by some works about the successful application of deep fully convolutional network (FCN) in computer vision, we design a perceptual loss to measure the error. It is now clear that the lower layers in the FCN usually will extract some general features shared by all objects like the horizontal (vertical) curves, while the higher layers are usually more objects oriented. This gives people the insight to train the network using different layers. Johnson then proposed the perceptual loss [29] which is the combination of the MSE of selected layers of the VGG model [51]. The authors claim in their paper that the early layers tends to produce images that are visually indistinguishable from the input; however if reconstruct from higher layers, image content and overall spatial structure are preserved but color, texture, and exact shape are not.

We will adopt the perceptual loss idea and design an adversary network to compute an additional reconstruction loss. The network structure can be seen in Figure 6.

The adversary net is fully convolutional with input and output both pre-computed multiscale basis. The network has an auto encoder structure and is pre-trained; i.e., we are going to solve the following minimization problem:

$$\min_{\theta_A} \frac{1}{m} \sum_i \|f(\phi_i) - \phi_i\|_2^2, \tag{14}$$

where ϕ_i is the multiscale basis and f is the adversary net associated with trainable parameter θ_A. Denote $f_j(\cdot)$ as the output of layer j of the adversary network. The additional reconstruction loss is then redefined as:

$$A(\theta_F, \theta_G) = \frac{1}{m} \sum_{i=1}^m \sum_{j \in I} \|f_j(G(F(\kappa_i))) - f_j(\phi_i)\|_2^2, \tag{15}$$

where I is the index set which contains some layers of the adversary net. The complete optimization problem can be now formulated as:

$$\min_{\theta_G, \theta_F} \lambda_1 C + \lambda_2 R + \lambda_3 A. \tag{16}$$

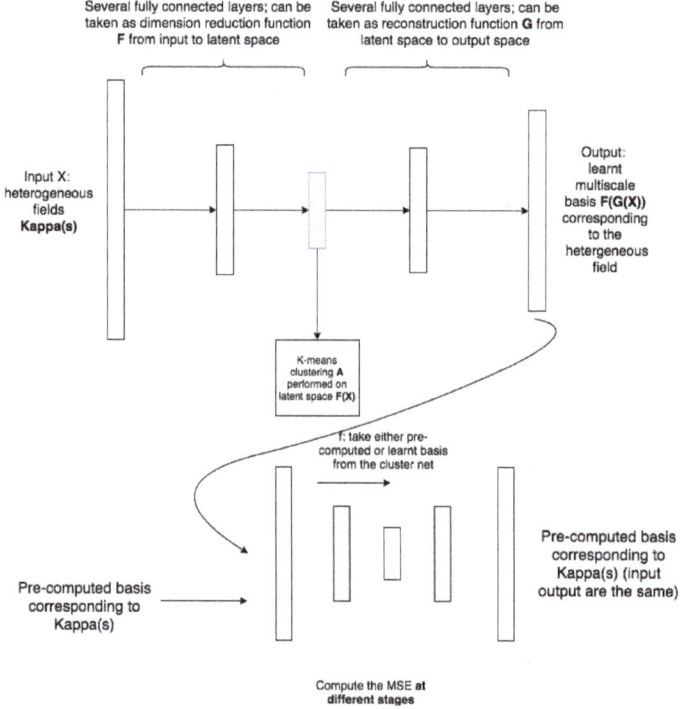

Figure 6. The complete network.

4. Numerical Experiments

In this section, we will demonstrate a series of experiments. We are going to apply our method on problems with high contrast including moving background and moving channels. he experiments are related to subsurface simulations. The moving background and moving channels simulate some important characteristics in the field. We numerically generate heterogeneous fields which contain moving channels and varying well rates. In Section 4.1, we first demonstrate a set of simulated heterogeneous oil fields to be used to train and solve the PDE modeling the reservoirs simulation; the deep learning model settings are also detailed in this section. In Section 4.2, we simulate some other more complicated heterogeneous fields and conduct the experiments to demonstrate the power of clustering algorithm. This experiments can show that our method is robust to handle complicated problems. In the last section, we will solve the PDE using the proposed method based on the heterogeneous field proposed in Section 4.1 and compute the relative error to demonstrate the accuracy of our method.

4.1. High Contrast Heterogeneous Fields with Moving Channels

We consider solving Equations (1)–(2) for a heterogeneous field with moving channels and changing background. Let us denote the heterogeneous field as $\kappa(x)$, where $x \in [0,1]^2$, then $\kappa(x) = 1000$ if x is in some channels which will be illustrated later and otherwise,

$$\kappa(x) = e^{\eta \cdot \sin(7\pi x)\sin(8\pi y) + \sin(10\pi x)\sin(12\pi y)},$$

where η follows discrete uniform distribution in $[0,1]$. The channels are moving and we include cases of the intersection of two channels and formation and dissipation of the channels in the fields. These simulate the realistic petroleum oil fields. In Figure 7, we demonstrate 20 heterogeneous fields.

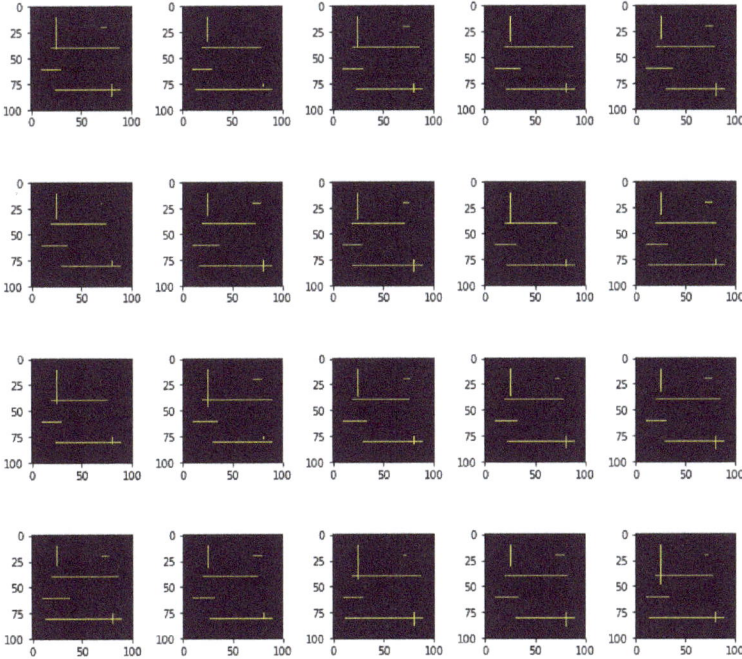

Figure 7. Heterogeneous fields, the yellow strips are the channels.

It can be observed from the images that, vertical channel (at around $x = 30$) (not always) intersects with horizontal channels (at around $y = 40$); and the channel at $x = 75, y = 25$ demonstrates the case of generation and degeneration of a channel.

We train the network using 600 samples using the Adam gradient descent. We find that the cluster assignment of 600 realizations in uncertain space is stable(fixed) when the gradient descent epoch reaches a certain number, so we set the stopping criteria to be: the assignment does not change for 100 iteration epochs; and the maximum number of iteration epochs is set to be 1000. We also find that the coefficients in Equation (16) can affect the training result. We set $\lambda_1 = \lambda_2 = \lambda_3 = 1$.

It should be noted that we train the network locally in each coarse neighborhood. The fine mesh element has size $1/100$ and 5 fine elements are merged into one coarse element.

The dimension reduction network F contains 4 fully connected layers to reduce the size of local coarse elements from 100 to $60, 40, 30, 20$ gradually. The K-means clustering is conducted in space $F(x)$ of dimension 20; the reconstruction net G is designed symmetrically with the reduction network F. The adversary net is fully convolutional. All convolution layers except the last layer have kernels of size 3 by 3 with stride 1; we use 1 by 1 convolution in the last layer to reduce the number of channels to 1. The number of channels is doubled if the spatial dimension is reduced and half-ed if the spatial dimension is increased. Max pooling of size 2 by 2 is used to reduce the spatial dimension in the encoder; and to increase the dimension in the decoder, we perform the nearest neighbor resize followed by convolution [52].

4.2. Results

We will present the numerical results of the proposed method in this section. We are going to show the cluster assignment experiment first, followed by two other experiments which demonstrate the error of the method.

4.2.1. Cluster Assignment in a Local Coarse Element

Before diving into the error analysis, we will show some of the cluster results in a local neighborhood. In this neighborhood, we manually created the cases such as: the extraction of a channel (longer), the expansion of a channel(wider), the discontinuity of a channel, the diagonal channels, the intersection of channels, and so on. In Figure 8, the number on top of each image is the cluster assignment ID number.

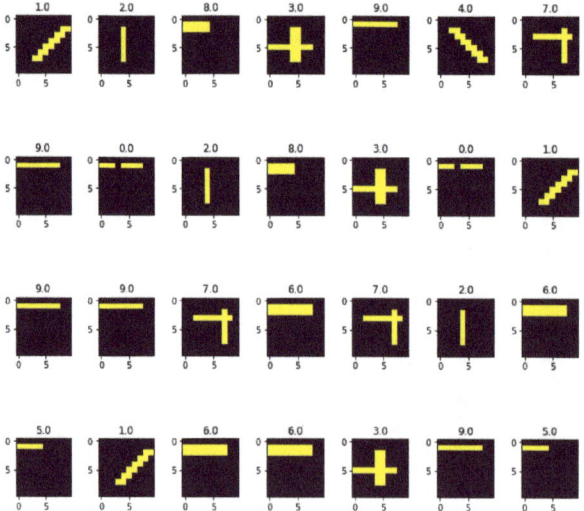

Figure 8. Cluster results of 28 samples, images shown are heterogeneous fields, the number on top of each image is the cluster assignment ID number.

We also demonstrate the clustering result in Figure 9 of another neighborhood which is around $(25, 45)$ in Figure 7. From the results in both Figures 8 and 9, we observe that our proposed clustering algorithm based on deep learning is able create a good clustering of the parameter space. That is, heterogeneous coefficients with similar spatial structures are grouped in the same cluster.

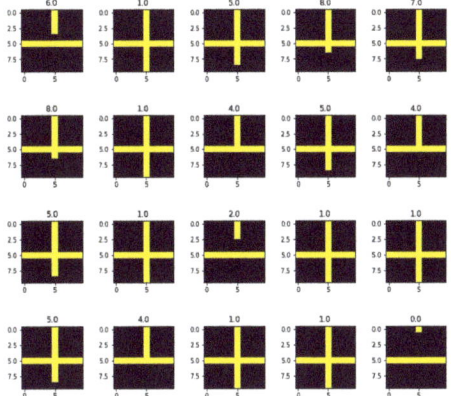

Figure 9. Cluster results of 20 samples, images shown are heterogeneous fields, the number on top of each image is the cluster assignment ID number.

4.2.2. Relation of Error and the Number of Clusters

In this section, we will demonstrate the error change when the hyperparamter—the number of clusters—increases. Given a new realization $\kappa(x, \hat{s})$ where $\hat{s}$ denotes the parameter and a fixed neighborhood, suppose the neighborhood of this realization will be fitted into cluster S_i by the model trained. We compute $\bar{\kappa}_i = \frac{1}{|S_i|} \sum_{j=1}^{|S_i|} \kappa_{ij}$ where $|S_i|$ is the number of points in this cluster S_i. The GMsFEM basis of this neighborhood can then be derived using $\bar{\kappa}_i$. We finally construct the solution using the GMsFEM basis pre-computed in all neighborhoods. We define the l_2 relative error as :

$$ratio = \frac{\int_D (u - u_H)^2 dx}{\int_D u^2 dx}, \qquad (17)$$

where u is the exact solution computed by finite element method with fine enough mesh and u_H is the solution of the proposed method. We test the network on newly generated 300 samples and take the average of the errors.

In this experiment, we calculate the l_2 relative error with the number of clusters increases. The number of clusters ranges from 5 to 11; and for each case, we will train the model and compute the l_2 relative error. The result can be seen in Figure 10 and it can be observed from the picture that, the error is decreasing with the number of cluster increases.

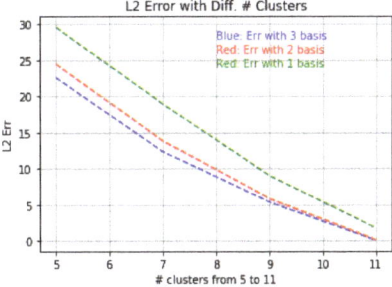

Figure 10. The l_2 error when the number of clusters changes, colors represent the number of GMsFEM basis.

4.2.3. Comparison of Cluster-based Method with Tradition Method

In the second experiments, we first compute the l_2 relative error (defined in Equation (17) with u_H denoting the GMsFEM solution) of traditional GMsFEM method with given $\kappa(x, \hat{s})$. This means that the construct multiscale basis functions using the particular realization $\kappa(x, \hat{s})$. We then compare this error with the cluster method proposed (11 clusters). The comparison can be seen in Figure 11.

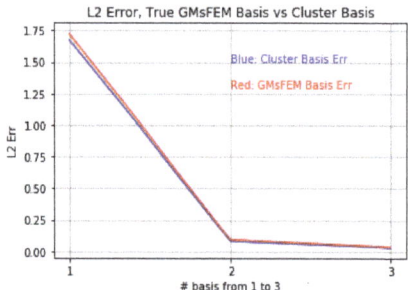

Figure 11. The l_2 error cluster solution (11 clusters) vs. solution by real $\kappa(x, \hat{s})$. Color represents number of basis.

It can be seen that the difference is negligible when the number of clusters reaches 11. We can then benefit from the deep learning; i.e., the fitting of $\kappa(x, \hat{s})$ into a cluster is fast; and since we will use the pre-computed basis, we also save time on computing the GMsFEM basis.

4.3. Effect of the Adversary Net

The target of this task is not the learning of multiscale basis; the multiscale basis in this work is just a supervision of learning the cluster. However, to demonstrate the effectiveness of the adversary network, we also test the the effect of the adversary net. There are many hyper-parameters like the number of clusters and coefficients of the loss function which can affect the result; so to reduce the influence from the clustering, we remove the clustering loss from the training, so this is a generative task which will generate the multiscale basis from the output of the first network in Figure 6. The loss function now can be defined as:

$$\min_{\theta_G, \theta_F} \lambda_1 R + \lambda_2 A, \quad (18)$$

where R and A are defined in Equations (12) and (15), separately; and λ_1 and λ_1 are both set to be 1. We compute the relative error with Equation (17) first by using the learned multiscale basis which is trained by Equation (18); and second by using the multiscale basis trained without the adversary loss Equation (15), i.e.,

$$\min_{\theta_G, \theta_F} A. \quad (19)$$

The l_2 relative error improves from 41.120439 to 36.760918 if we add one middle layer from the adversary net.

We also calculate the MSE difference of two learned basis (by loss Equation (18) and Equation (19), separately) and real multiscale basis, i.e., we calculate $\|B_{\text{learned basis}} - B_{\text{real basis}}\|_{MSE}$, where $B_{\text{learned basis}}$ refers to two basis trained with Equation (18) and Equation (19), separately and $B_{\text{real basis}}$ is the real multiscale basis formed using the input heterogeneous field. The MSE amazingly decreases from 0.9073400 to 0.748312 if we use basis trained with the adversary loss Equation (18). This can show the benefit from the adversary net.

5. Conclusions

We propose a deep learning clustering technique within GMsFEM to solve flows in heterogeneous media. The main idea is to cluster the uncertainty space such that we can reduce the number of multiscale basis functions for each coarse block across the uncertainty space. We propose the adversary loss motivated by the perceptual loss in the computer vision task. We use convolutional neural networks combined with some techniques in adversary neural networks, where the loss function is composed of several parts that includes terms related to clusters and reconstruction of basis functions. We present numerical results for channelized permeability fields in the examples of flows in porous media. In future, we would like to study the relation between convolutional layers and quantities related to multiscale basis functions. In addition, we are going to study the application of our method in the area of multiscale social network and other studies like extreme value prediction.

Author Contributions: All authors have contributed to methodology and validation. Simulations are performed by Z.Z. All authors have read and agreed to the published version of the manuscript.

Funding: Eric Chung's work is partially supported by the Hong Kong RGC General Research Fund (Project numbers 14304719 and 14302018) and the CUHK Faculty of Science Direct Grant 2018-19. Y.E. would like to thank the partial support from NSF 1620318 and NSF Tripod 1934904. Y.E. would also like to acknowledge the support of Mega-grant of the Russian Federation Government (N 14.Y26.31.0013).

Conflicts of Interest: The authors declare no conflict of interest.

References

1. Tsiropoulou, E.; Koukas, K.; Papavassiliou, S. A socio-physical and mobility-aware coalition formation mechanism in public safety networks. *EAI Endorsed Trans. Future Internet* **2018**, *4*, 154176. [CrossRef]
2. Thai, M.T.; Wu, W.; Xiong, H. *Big Data in Complex and Social Networks*; CRC Press: Boca Raton, FL, USA, 2016.
3. Calo, V.M.; Efendiev, Y.; Galvis, J.; Ghommem, M. Multiscale empirical interpolation for solving nonlinear PDEs. *J. Comput. Phys.* **2014**, *278*, 204–220. [CrossRef]
4. Chung, E.T.; Efendiev, Y.; Leung, W.T. Residual-driven online generalized multiscale finite element methods. *J. Comput. Phys.* **2015**, *302*, 176–190. [CrossRef]
5. Chung, E.T.; Efendiev, Y.; Leung, W.T. An online generalized multiscale discontinuous Galerkin method (GMsDGM) for flows in heterogeneous media. *Commun. Comput. Phys.* **2017**, *21*, 401–422. [CrossRef]
6. Chung, E.T.; Efendiev, Y.; Li, G. An adaptive GMsFEM for high-contrast flow problems. *J. Comput. Phys.* **2014**, *273*, 54–76. [CrossRef]
7. Chung, E.T.; Efendiev, Y.; Li, G.; Vasilyeva, M. Generalized multiscale finite element methods for problems in perforated heterogeneous domains. *Appl. Anal.* **2016**, *95*, 2254–2279. [CrossRef]
8. Efendiev, Y.; Galvis, J.; Lazarov, R.; Moon, M.; Sarkis, M. Generalized multiscale finite element method. Symmetric interior penalty coupling. *J. Comput. Phys.* **2013**, *255*, 1–15. [CrossRef]
9. Efendiev, Y.; Galvis, J.; Li, G.; Presho, M. Generalized multiscale finite element methods: Oversampling strategies. *Int. J. Multiscale Comput. Eng.* **2014**, *12*, 465–484. [CrossRef]
10. Chung, E.; Efendiev, Y.; Hou, T.Y. Adaptive multiscale model reduction with generalized multiscale finite element methods. *J. Comput. Phys.* **2016**, *320*, 69–95. [CrossRef]
11. Chung, E.; Efendiev, Y.; Fu, S. Generalized multiscale finite element method for elasticity equations. *Int. J. Geomath.* **2014**, *5*, 225–254. [CrossRef]
12. Chung, E.; Vasilyeva, M.; Wang, Y. A conservative local multiscale model reduction technique for Stokes flows in heterogeneous perforated domains. *J. Comput. Appl. Math.* **2017**, *321*, 389–405. [CrossRef]
13. Chung, E.T.; Efendiev, Y.; Leung, W.T.; Zhang, Z. Cluster-based generalized multiscale finite element method for elliptic PDEs with random coefficients. *J. Comput. Phys.* **2018**, *371*, 606–617. [CrossRef]
14. Karhunen, K. *Über Lineare Methoden in der Wahrscheinlichkeitsrechnung*; Suomalainen Tiedeakatemia: Helsinki, Finland 1947; Volume 37.
15. Lloyd, S. Least squares quantization in PCM. *IEEE Trans. Inf. Theory* **1982**, *28*, 129–137. [CrossRef]
16. Wang, Y.; Cheung, S.W.; Chung, E.T.; Efendiev, Y.; Wang, M. Deep multiscale model learning. *J. Comput. Phys.* **2020**, *406*, 109071. [CrossRef]

17. Wang, M.; Cheung, S.W.; Leung, W.T.; Chung, E.T.; Efendiev, Y.; Wheeler, M. Reduced-order deep learning for flow dynamics. The interplay between deep learning and model reduction. *J. Comput. Phys.* **2020**, *401*, 108939. [CrossRef]
18. Vasilyeva, M.; Leung, W.T.; Chung, E.T.; Efendiev, Y.; Wheeler, M. Learning macroscopic parameters in nonlinear multiscale simulations using nonlocal multicontinua upscaling techniques. *arXiv* **2019**, arXiv:1907.02921.
19. Cheung, S.W.; Chung, E.T.; Efendiev, Y.; Gildin, E.; Wang, Y.; Zhang, J. Deep global model reduction learning in porous media flow simulation. *Comput. Geosci.* **2020**, *24*, 261–274. [CrossRef]
20. Wang, M.; Cheung, S.W.; Chung, E.T.; Efendiev, Y.; Leung, W.T.; Wang, Y. Prediction of discretization of gmsfem using deep learning. *Mathematics* **2019**, *7*, 412. [CrossRef]
21. Goodfellow, I.; Bengio, Y.; Courville, A. *Deep Learning*; MIT Press: Cambridge, MA, USA, 2016.
22. Caron, M.; Bojanowski, P.; Joulin, A.; Douze, M. Deep clustering for unsupervised learning of visual features. In Proceedings of the European Conference on Computer Vision (ECCV), Munich, Germany, 8–14 September 2018; pp. 132–149.
23. Xie, J.; Girshick, R.; Farhadi, A. Unsupervised deep embedding for clustering analysis. In Proceedings of the International Conference on Machine Learning, New York, NY, USA, 19–24 June 2016; pp. 478–487.
24. Yang, B.; Fu, X.; Sidiropoulos, N.D.; Hong, M. Towards k-means-friendly spaces: Simultaneous deep learning and clustering. In Proceedings of the 34th International Conference on Machine Learning-Volume 70. JMLR.org, Sydney, Australia, 6–11 August 2017; pp. 3861–3870.
25. Bishop, C.M. *Pattern Recognition and Machine Learning*; Springer: Berlin, Germany, 2006.
26. Isola, P.; Zhu, J.Y.; Zhou, T.; Efros, A.A. Image-to-image translation with conditional adversarial networks. In Proceedings of the IEEE Conference on Computer Vision and Pattern Recognition, Honolulu, HI, USA, 21–26 July 2017; pp. 1125–1134.
27. Ledig, C.; Theis, L.; Huszár, F.; Caballero, J.; Cunningham, A.; Acosta, A.; Aitken, A.; Tejani, A.; Totz, J.; Wang, Z.; et al. Photo-realistic single image super-resolution using a generative adversarial network. In Proceedings of the IEEE Conference on Computer Vision and Pattern Recognition, Honolulu, HI, USA, 21–26 July 2017; pp. 4681–4690.
28. Lai, W.S.; Huang, J.B.; Ahuja, N.; Yang, M.H. Deep laplacian pyramid networks for fast and accurate super-resolution. In Proceedings of the IEEE Conference on Computer Vision and Pattern Recognition, Honolulu, HI, USA, 21–26 July 2017; pp. 624–632.
29. Johnson, J.; Alahi, A.; Fei-Fei, L. Perceptual losses for real-time style transfer and super-resolution. In Proceedings of the European Conference on Computer Vision, Amsterdam, The Netherlands, 11–14 October 2016; Springer: Berlin, Germany, 2016; pp. 694–711.
30. Dong, C.; Loy, C.C.; He, K.; Tang, X. Image super-resolution using deep convolutional networks. *IEEE Trans. Pattern Anal. Mach. Intell.* **2015**, *38*, 295–307. [CrossRef]
31. Kim, J.; Kwon Lee, J.; Mu Lee, K. Accurate image super-resolution using very deep convolutional networks. In Proceedings of the IEEE Conference on Computer Vision and Pattern Recognition, Las Vegas, NV, USA, 26 June–1 July 2016; pp. 1646–1654.
32. Zhang, Y.; Li, K.; Li, K.; Wang, L.; Zhong, B.; Fu, Y. Image super-resolution using very deep residual channel attention networks. In Proceedings of the European Conference on Computer Vision (ECCV), Munich, Germany, 8–14 September 2018; pp. 286–301.
33. Zhang, Y.; Tian, Y.; Kong, Y.; Zhong, B.; Fu, Y. Residual dense network for image super-resolution. In Proceedings of the IEEE Conference on Computer Vision and Pattern Recognition, Salt Lake City, UT, USA, 18–22 June 2018; pp. 2472–2481.
34. Tai, Y.; Yang, J.; Liu, X. Image super-resolution via deep recursive residual network. In Proceedings of the IEEE Conference on Computer Vision and Pattern Recognition, Honolulu, HI, USA, 21–26 July 2017; pp. 3147–3155.
35. Lim, B.; Son, S.; Kim, H.; Nah, S.; Mu Lee, K. Enhanced deep residual networks for single image super-resolution. In Proceedings of the IEEE Conference on Computer Vision and Pattern Recognition Workshops, Honolulu, HI, USA, 21–26 July 2017; pp. 136–144.
36. Tai, Y.; Yang, J.; Liu, X.; Xu, C. Memnet: A persistent memory network for image restoration. In Proceedings of the IEEE International Conference on Computer Vision, Venice, Italy, 22–29 October 2017; pp. 4539–4547.

37. Goodfellow, I.; Pouget-Abadie, J.; Mirza, M.; Xu, B.; Warde-Farley, D.; Ozair, S.; Courville, A.; Bengio, Y. Generative adversarial nets. In Proceedings of the Advances in Neural Information Processing Systems, Montreal, QC, Canada, 8–13 December 2014; pp. 2672–2680.
38. Hu, J.; Shen, L.; Sun, G. Squeeze-and-excitation networks. In Proceedings of the IEEE Conference on Computer Vision and Pattern Recognition, Salt Lake City, UT, USA, 18–22 June 2018; pp. 7132–7141.
39. Zhang, H.; Goodfellow, I.; Metaxas, D.; Odena, A. Self-attention generative adversarial networks. *arXiv* **2018**, arXiv:1805.08318.
40. Long, J.; Shelhamer, E.; Darrell, T. Fully convolutional networks for semantic segmentation. In Proceedings of the IEEE Conference on Computer Vision and Pattern Recognition, Boston, MA, USA; 7–12 June 2015; pp. 3431–3440.
41. Szegedy, C.; Ioffe, S.; Vanhoucke, V.; Alemi, A.A. Inception-v4, inception-resnet and the impact of residual connections on learning. In Proceedings of the Thirty-First AAAI Conference on Artificial Intelligence, San Francisco, CA, USA, 4–9 February 2017.
42. Szegedy, C.; Vanhoucke, V.; Ioffe, S.; Shlens, J.; Wojna, Z. Rethinking the inception architecture for computer vision. In Proceedings of the IEEE Conference on Computer Vision and Pattern Recognition, Las Vegas, NV, USA, 26 June–1 July 2016; pp. 2818–2826.
43. Badrinarayanan, V.; Kendall, A.; Cipolla, R. Segnet: A deep convolutional encoder-decoder architecture for image segmentation. *IEEE Trans. Pattern Anal. Mach. Intell.* **2017**, *39*, 2481–2495. [CrossRef] [PubMed]
44. Noh, H.; Hong, S.; Han, B. Learning deconvolution network for semantic segmentation. In Proceedings of the IEEE International Conference on Computer Vision, Santiago, Chile, 11–18 December 2015; pp. 1520–1528.
45. Huang, G.; Liu, Z.; Van Der Maaten, L.; Weinberger, K.Q. Densely connected convolutional networks. In Proceedings of the IEEE Conference on Computer Vision and Pattern Recognition, Honolulu, HI, USA, 21–26 July 2017; pp. 4700–4708.
46. Efendiev, Y.; Galvis, J.; Hou, T.Y. Generalized multiscale finite element methods (GMsFEM). *J. Comput. Phys.* **2013**, *251*, 116–135. [CrossRef]
47. Hou, T.Y.; Wu, X.H. A multiscale finite element method for elliptic problems in composite materials and porous media. *J. Comput. Phys.* **1997**, *134*, 169–189. [CrossRef]
48. Jenny, P.; Lee, S.; Tchelepi, H.A. Multi-scale finite-volume method for elliptic problems in subsurface flow simulation. *J. Comput. Phys.* **2003**, *187*, 47–67. [CrossRef]
49. Calo, V.M.; Efendiev, Y.; Galvis, J.; Li, G. Randomized oversampling for generalized multiscale finite element methods. *Multiscale Model. Simul.* **2016**, *14*, 482–501. [CrossRef]
50. Vidal-Jordana, A.; Montalban, X. Multiple sclerosis: Epidemiologic, clinical, and therapeutic aspects. *Neuroimaging Clin.* **2017**, *27*, 195–204. [CrossRef]
51. Simonyan, K.; Zisserman, A. Very deep convolutional networks for large-scale image recognition. *arXiv* **2014**, arXiv:1409.1556.
52. Odena, A.; Dumoulin, V.; Olah, C. Deconvolution and checkerboard artifacts. *Distill* **2016**, *1*, e3. [CrossRef]

© 2020 by the authors. Licensee MDPI, Basel, Switzerland. This article is an open access article distributed under the terms and conditions of the Creative Commons Attribution (CC BY) license (http://creativecommons.org/licenses/by/4.0/).

Article

Angular Correlation Using Rogers-Szegő-Chaos

Christine Schmid [1,*] and Kyle J. DeMars [2]

[1] Department of Mechanical and Aerospace Engineering, Missouri University of Science and Technology, Rolla, MO 65409, USA
[2] Department of Aerospace Engineering, Texas A&M University, College Station, TX 77843, USA; demars@tamu.edu
* Correspondence: clshg6@mst.edu

Received: 30 December 2019; Accepted: 21 January 2020; Published: 1 February 2020

Abstract: Polynomial chaos expresses a probability density function (pdf) as a linear combination of basis polynomials. If the density and basis polynomials are over the same field, any set of basis polynomials can describe the pdf; however, the most logical choice of polynomials is the family that is orthogonal with respect to the pdf. This problem is well-studied over the field of real numbers and has been shown to be valid for the complex unit circle in one dimension. The current framework for circular polynomial chaos is extended to multiple angular dimensions with the inclusion of correlation terms. Uncertainty propagation of heading angle and angular velocity is investigated using polynomial chaos and compared against Monte Carlo simulation.

Keywords: polynomial chaos; Szegő polynomials; directional statistics; Rogers-Szegő; state estimation

1. Introduction

Engineering is an imperfect science. Noisy measurements from sensors in state estimation [1,2], a constantly changing environment in guidance [3–5], and improperly actuated controls [6] are all major sources of error. The more these sources of error are understood, the better the final product will be. Ideally, every variable with some sort of uncertainty associated with it would be completely and analytically described with its probability density function (pdf). Unfortunately, even if this is feasible for the initialization of a random variable, its evolution through time rarely yields a pdf with an analytic form. If the pdf cannot be given in analytic form, then approximations and assumptions must be made.

In many cases, a random variable is quantified using only its first two moments—as with the unscented transform [7]—and a further assumption is that the distribution is Gaussian. In cases where the variable's uncertainty is relatively small and the dynamics governing its evolution are not highly nonlinear, this is not necessarily a poor assumption. In these cases, the higher order moments are highly dependent on the first two moments; i.e., there is a minimal amount of unique information in the higher order moments. In contrast, if either the uncertainty is large or the dynamics become highly nonlinear, the higher order moments become less dependent on the first two moments and contain larger amounts of unique information. In this case, the error associated with using only the first two moments becomes significant [8,9].

One method of quantifying uncertainty that does not require an assumption of the random variable's pdf is polynomial chaos expansion (PCE) [10–14]. PCE characterizes a random variable as a coordinate in a polynomial vector space. Useful deterministic information about the random variable lies in this coordinate, including the moments of the random variable [15]. The expression of the coordinate depends on the basis in which it is expressed. In the case of PCE, the bases are made up of polynomials that are chosen based on the assumed density of the random variable; however, any random variable can be represented using any basis [16]. It is strongly noted that assuming the density of the random variable simply eases computation; with enough computing power, any random variable can be quantified with any basis [17]. The most common basis polynomials are those that are orthogonal with respect to common pdfs, such as the Hermite-Gaussian and Jacobi-beta polynomial-pdf pairs.

Polynomial chaos has been used to quantify and propagate the uncertainty in nonlinear systems including fluid dynamics [18–21], orbital mechanics in multiple element spaces [22], and has been expanded to facilitate Gaussian mixture models [23]. While polynomial chaos has been well-studied for variables that exist in the n-dimensional space of real numbers ($\mathbb{R}^n$), many variables do not lie in this field. The variables that are of concern in this paper are angular variables. When estimating an angular random variable, such as the true anomaly of a body in orbit or vehicle attitude, a common approach is to assume that the variable is approximately equal to its projection on $\mathbb{R}^n$ and use methods designed for variables on that space. When the uncertainty is very small, this approximation is relatively effective; however, as the uncertainty increases, this approximation becomes invalid. Recently, work on directional statistics has been incorporated into state estimation, providing a method for estimating angular random variables directly on the n-dimensional special orthogonal group ($\mathbb{S}^n$) [24–26]. Recent work [27] has shown that polynomial chaos can be used to estimate the mean and variance of one-dimensional angular random variables and the set of equinoctial orbital elements elements [28]. In neither case was a correlation estimated that included an angular state.

The sections of this paper include a detailed discussion of orthogonal polynomials in Section 2, including those that are orthogonal on the unit circle and two of the more common circular pdfs. In Section 3 an overview of polynomial chaos is given as well as the extension that has been made to include angular random variables, including the correlation between two angles. Finally, in Section 4, numerical results are presented comparing the correlation between two angular random variables calculated using Monte Carlo and PCE.

2. Orthogonal Polynomials

Let $\mathbb{V}^n$ be an n-dimensional vector space. A basis of this vector space is the minimal set of vectors that spans the vector space. An orthogonal basis is a subset of bases consisting of exactly n basis vectors such that the inner product between any two basis vectors, β_m and β_n, is proportional to the Kronecker delta (δ_{mn}). Given mathematically with angle brackets, this orthogonal inner product takes the following form:

$$\langle \beta_m, \beta_n \rangle = c\, \delta_{mn},$$

where m and n are part of the set of positive integers, including zero. In the event $c = 1$, the set is termed orthonormal. The ith standard basis vector of $\mathbb{V}^n$ is generally the ith vector of the n-dimensional identity; however, there are infinitely many bases for each vector space. It should be noted that it is not a requirement that a basis be orthogonal, merely linearly independent; however, the use of non-orthogonal bases is practically unheard-of.

An element $\alpha \in \mathbb{V}$ can be expressed in terms of an ordered basis $\mathcal{B} = \{\beta_1, \beta_2, \ldots, \beta_n\}$, as the linear combination

$$\alpha = a_1 \beta_1 + a_2 \beta_2 + \cdots + a_n \beta_n, \tag{1}$$

where $[a_1, a_2, \ldots, a_n]$ is the coordinate of α. While any set of independent vectors can be used as a basis, different bases can prove beneficial—possibly by making the system more intuitive or more mathematically straightforward. When expressing the state of a physical system, the selection of a coordinate frame is effectively choosing a basis for the inhabited vector space. Consider a satellite in orbit. If the satellite's ground track is of high importance (such as weather or telecommunications satellites), an Earth-fixed frame would be ideal. However, in cases where a satellite's actions are dictated by other space-based objects (such as proximity operations), a body-fixed frame would be ideal.

It is common to constrict the term vector space to the spaces that are easiest to visualize, most notably a Cartesian space, where the bases are vectors radiating from the origin at right angles. The term vector space is much more broad than this though. A vector space need only contain the zero element and be closed under both scalar addition and multiplication, which applies to much more than vectors.

Most notable in this work is the idea of a polynomial vector space. Let $\mathbb{P}^{n+1}$ be an $(n+1)$-dimensional vector space made up of all polynomials of positive degree n or less with standard basis $\mathcal{B} = \{1, x, \ldots, x^n\}$. The inner product with respect to the function ω on the real-valued polynomial space is given by

$$\langle f(x), g(x) \rangle_{\omega(x)} = \int_{\mathbb{S}} f(x) g(x) d\omega(x),$$

where $\omega(x)$ is a non-decreasing function with support $\mathbb{S}$ and f and g are any two polynomials of degree n or less. A polynomial family $\Phi(x)$ is a set of polynomials with monotonically increasing order that are orthogonal. The orthogonality condition is given mathematically as

$$\langle \phi_m(x), \phi_n(x) \rangle_{\omega(x)} = \int_{\mathbb{S}} \phi_m(x) \phi_n(x) d\omega(x) = 0 \tag{2a}$$

$$\langle \phi_m^2(x) \rangle_{\omega(x)} = \int_{\mathbb{S}} \phi_m(x) \phi_m(x) d\omega(x) = c, \tag{2b}$$

where $\phi_k(x)$ is the polynomial of order k, c is a constant, and $\mathbb{S}$ is the support of the non-decreasing function $\omega(x)$. Note that while polynomials of negative orders ($k < 0$), referred to as Laurent polynomials, exist, they are not covered in this work.

The most commonly used polynomial families are categorized in the Askey scheme, which groups the polynomials based on the generalized hypergeometric function ($_pF_q$) from which they are generated [29–31]. Table 1 lists some of the polynomial families, their support, the non-decreasing function they are orthogonal with respect to (commonly referred to as a weight function), and the hypergeometric function they can be written in terms of. For completeness, Table 1 lists both continuous and discrete polynomial groups; however, the remainder of this work only considers continuous polynomials.

Table 1. Common Orthogonal Polynomials.

Type	Polynomial	Hypergeometric Series	Support	Weight Function/Distribution
Continuous	Legendre	$_2F_1$	$[-1, 1]$	Uniform
	Jacobi	$_2F_1$	$[-1, 1]$	Beta
	Laguerre	$_1F_1$	$[0, \infty)$	Exponential
	Probabilists' Hermite	$_2F_0$	$(-\infty, \infty)$	Normal
Discrete	Charlier	$_2F_0$	$\{0, 1, 2, \ldots\}$	Poisson
	Meixner	$_2F_1$	$\{0, 1, 2, \ldots\}$	Negative Binomial
	Krawtchouk	$_2F_1$	$\{0, 1, \ldots, N\}$	Binomial
	Hahn	$_3F_2$	$\{0, 1, \ldots, N\}$	Hypergeometric

2.1. Polynomials Orthogonal on the Unit Circle

While the Askey polynomials are useful in many applications, their standard forms place them in the polynomial ring $\mathbb{R}[x]$, or all polynomials with real-valued coefficients that are closed under polynomial addition and multiplication. These polynomials are orthogonal with respect to measures on the real line. In the event that a set of polynomials orthogonal with respect a measure on a curved interval (e.g., the unit circle) is desired, the Askey polynomials would be insufficient.

In [32], Szegő uses the connection between points on the unit circle and points on a finite real interval to develop polynomials that are orthogonal on the unit circle. Polynomials of this type are now known as Szegő polynomials. Since the unit circle is defined to have unit radius, every point can be described on a real interval of length 2π and mapped to the complex variable $\vartheta = e^{i\theta}$, where i is the imaginary unit. All use of the variable ϑ in the following corresponds to this definition. The orthogonality expression for the Szegő polynomials, $\phi_k(\vartheta)$, is

$$\langle \phi_m(\vartheta), \phi_n(\vartheta) \rangle_{\omega(\theta)} = \frac{1}{2\pi} \int_{-\pi}^{\pi} \phi_m(\vartheta) \overline{\phi_n(\vartheta)} \omega(\theta) d\theta = \delta_{mn},$$

where $\overline{\phi_n(\vartheta)}$ is the complex conjugate of $\phi_n(\vartheta)$ and $\omega(\theta)$ is the monotonically increasing weight function over the support. Note that, as opposed to Equation (2a), the Kronecker delta is not scaled, implying all polynomials using Szegő's formulation are orthonormal.

While the general formulation outlined by Szegő is cumbersome—requiring the calculation of Fourier coefficients corresponding to the weight function and large matrix determinants—it does provide a framework for developing a set of polynomials orthogonal with respect to any conceivable continuous weight function. In addition to the initial research done by Szegő, further studies have investigated polynomials orthogonal on the unit circle [33–38].

Fortunately, there exist some polynomial families that are given explicitly, such as the Rogers-Szegő polynomials. The Rogers-Szegő polynomials have been well-studied [39–41] and were developed by Szegő based on work done by Rogers over the q-Hermite polynomials. For a more detailed description of the relationship between the Askey scheme of polynomials and their q-analog, the reader is encouraged to reference [31,42].

The generating function for the Rogers-Szegő polynomials is given as

$$\phi_n(\vartheta; q) = \sum_{k=0}^{n} \binom{n}{k}_q \vartheta^k, \qquad (3)$$

where $\binom{n}{k}_q$ is the q-binomial

$$\binom{n}{k}_q = \frac{(q;q)_n}{(q;q)_k (q;q)_{n-k}} \qquad \text{and} \qquad (a;q)_n = \prod_{j=0}^{n-1}(1 - aq^j).$$

The weight function that satisfies the orthogonality condition of these polynomials is

$$\omega(\theta) = \frac{1}{\sqrt{-2\pi \ln(q)}} \sum_{j=-\infty}^{\infty} \exp\left\{ \frac{(\theta + 2\pi j)^2}{2 \ln(q)} \right\} \qquad 0 < q < 1. \qquad (4)$$

In addition to the generating function, a three-step recurrence [43] exists, which is given by

$$\phi_{n+1}(\vartheta; q) = (1 + \vartheta)\phi_n(\vartheta; q) - (1 - q^n)\vartheta \phi_{n-1}(\vartheta; q). \qquad (5)$$

For convenience, the first five polynomials are:

$\phi_0 = 1$
$\phi_1 = \vartheta + 1$
$\phi_2 = \vartheta^2 + (q+1)\vartheta + 1$
$\phi_3 = \vartheta^3 + (q^2 + q + 1)\vartheta^2 + (q^2 + q + 1)\vartheta + 1$
$\phi_4 = \vartheta^4 + (q+1)(q^2 + 1)\vartheta^3 + (q+1)(q^2 + q + 1)\vartheta^2 + (q+1)(q^2 + 1)\vartheta + 1.$

As is apparent, the q-binomial term causes the coefficients to be symmetric, which eases computation, and additionally, the polynomials are naturally monic.

2.2. Distributions on the Unit Circle

With the formulation of polynomials orthogonal on the unit circle, the weight function $\omega(\theta)$ has been continuously mentioned but not specifically addressed. In the general case, the weight function can be any non-decreasing function; however, the most common polynomial families are those that are orthogonal with respect to well-known pdfs, such as the ones listed in Table 1. Because weight functions must exist over the same support as the corresponding polynomials, pdfs over the unit circle are required for polynomial orthogonal on the unit circle.

2.2.1. Von Mises Distribution

One of the most common distributions used in directional statistics is the von Mises/von Mises-Fisher distribution [44–46]. The von Mises distribution lies on $\mathbb{S}^1$ (the subspace of $\mathbb{R}^2$ containing all points that are unit distance from the origin), whereas the von Mises-Fisher distribution has extensions into higher dimensional spheres. The circular von Mises pdf is given as [24]

$$p_m(\theta; \mu, \kappa) = \frac{e^{\kappa \cos(\theta - \mu)}}{2\pi I_0(\kappa)},$$

where μ is the mean angular direction on a 2π interval (usually $[-\pi, \pi]$), $\kappa \geq 0$ is a concentration parameter (similar to the inverse of the standard deviation), and I_0 is the zeroth order modified Bessel function of the first kind. The reason this distribution is so common is its close similarity to the normal distribution. This can be seen in Figure 1a, where von Mises distributions of various concentration parameters are plotted.

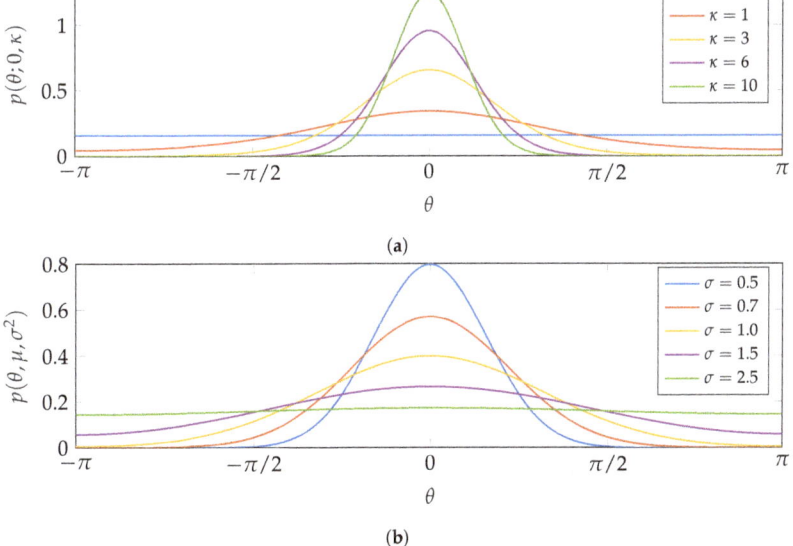

Figure 1. Common Circular probability density functions (pdfs). (**a**) Circular von Mises distribution for multiple values of κ. (**b**) Wrapped normal distribution for multiple values of σ.

2.2.2. Wrapped Distributions

The easiest to visualize circular distribution, or rather group of distributions, that is discussed is the set of wrapped distributions. The wrapped distributions take a distribution on the real line and wrap it onto the unit circle according to

$$p_w(\theta) = \sum_{j=-\infty}^{\infty} p(\theta + 2\pi j),$$

where the support of $p(\cdot)$ is an interval of $\mathbb{R}$, and the domain of p_w is an interval on $\mathbb{R}$ with length 2π. For example, wrapping a normal distribution takes the pdf

$$p_n(x; \mu, \sigma^2) = \frac{1}{\sqrt{2\pi\sigma^2}} \exp\left\{-\frac{(x-\mu)^2}{2\sigma^2}\right\},$$

where the domain of x is $\mathbb{R}$, and μ and σ are the mean and standard deviation, respectively, and wraps it, resulting in the wrapped pdf (in this case wrapped normal)

$$p_{wn}(\theta; \mu, \sigma^2) = \frac{1}{\sqrt{2\pi\sigma^2}} \sum_{j=-\infty}^{\infty} \exp\left\{-\frac{(\theta - \mu + 2\pi j)^2}{2\sigma^2}\right\},$$

where the domain of θ is an interval on $\mathbb{R}$ with length 2π. Zero-mean normal distributions with varying values of σ are wrapped, with the results shown in Figure 1b.

Recall the weight function of the Rogers-Szegő polynomials in Equation (4). As the log function is monotonically increasing, the term $\log(1/q)$ increases monotonically as q decreases. Observing the extremes of q: as q approaches 1, $\log(1/q)$ approaches 0, and as q approaches 0, $\log(1/q)$ approaches ∞. Letting $\log(1/q) = \sigma^2$ and $\mu = 0$, this becomes a zero-mean wrapped normal distribution.

It is clear from Figure 1 that both distributions described previously have strong similarities to the unwrapped normal distribution. Figure 1 also shows the difference in the standard deviation parameter. Whereas the wrapped normal distribution directly uses the standard deviation of the unwrapped distribution, the von Mises distribution is with respect to concentration parameter that is inversely related to the dispersion of the random variable. This makes the wrapped normal distribution slightly more intuitive when comparing with an unwrapped normal distribution.

2.2.3. Directional Statistics

The estimation of stochastic variables generally relies on calculating the statistics of that variable. Most notable of these statistics are the mean and variance, the first two central moments. For pdfs ($p(x)$) on the real line that are continuously integrable, the central moments are given as

$$\begin{aligned} \mu_1 &= \int_\mathbb{S} x p(x) dx \\ \mu_k &= \int_\mathbb{S} (x - \mu_1)^k p(x) dx \quad k \geq 2, \end{aligned} \tag{6}$$

where $\mathbb{S}$ is the support of $p(x)$. Although less utilized in general, raw moments are commonly used in directional statistics and are given as

$$\rho_k = \int_\mathbb{S} x^k p(x) dx,$$

or

$$\rho_k = \int_\mathbb{S} \theta^k p(\theta) d\theta,$$

where the slight distinction is that the integration variable is within the variable $\vartheta = e^{i\theta}$. In addition, mean direction and circular variance are not the first and second central moments [24]. Instead, both are calculated from the first moment's angle (θ_1) and length (R_1):

$$\theta_1 = \arg(\rho_1) = \tan^{-1}\frac{\mathrm{imag}(\rho_1)}{\mathrm{real}(\rho_1)} \tag{7a}$$

$$R_1 = \|\rho_1\|, \tag{7b}$$

where $\|\cdot\|$ is the l^2-norm. From Mardia [24], the length can be used to calculate the circular variance V_1 and circular standard deviation σ_1 according to

$$V_1 = 1 - R_1 \tag{8a}$$

$$\sigma_1 = \sqrt{-2\ln(R_1)}. \tag{8b}$$

Effectively, as the length of the moment decreases, the concentration of the pdf about the mean direction decreases and the unwrapped standard deviation (USTD) increases. Note that while the subscript in Equations (7) and (8) is 1, there are corresponding mean directions and lengths associated with all moments; however, these are rarely used in applications.

3. Polynomial Chaos

At any given instance in time, the deviation of the estimate from the truth can be approximated as a Gaussian distribution centered at the mean of the estimate. The space of these mean-centered Gaussians is known as a Gaussian linear space [11]; when that space is closed (i.e., the distributions have finite second moments), it falls into the Gaussian Hilbert space $\mathcal{H}$. At this point, what is needed is a way to quantify $\mathcal{H}$, as this gives the uncertainty between the estimate and the truth. This can be achieved by projecting $\mathcal{H}$ onto a complete set of orthogonal polynomials when those basis functions are evaluated at a random variable $\xi \in \mathcal{H}$. While the distribution at any point in time natively exists in $\mathcal{H}$, its projection onto the set of orthogonal polynomials provides a way of quantifying it by means of the ordered coordinates, as in Equation (1).

The homogeneous chaos [10] specifies ξ to be normally distributed with zero mean and unit variance (i.e., unit Gaussian), and the orthogonal polynomials to be the Hermite polynomials due to their orthogonality with respect to the standard Gaussian pdf [47]. Not only does this apply for Gaussian processes, but the Cameron-Martin theorem [48] says that this applies for any process with a finite second moment. Although the solution does converge as the number of orthogonal polynomials increases, further development has shown that, for different stochastic processes, certain basis functions cause the solution to converge faster [16], leading to the more general polynomial chaos (PC).

To begin applying this method mathematically for a general stochastic process, let a stochastic variable, ε, be expressed as the linear combination over an infinite-dimensional vector space, i.e.,

$$\varepsilon(x, \xi) = \sum_{k=0}^{\infty} \epsilon_k(x) \Psi_k(\xi), \tag{9}$$

where $\epsilon_i(x)$ is the deterministic component and $\Psi_i(\xi)$ is an ith-order orthogonal basis function evaluated at, and orthogonal with respect to, the weight function, ξ. The polynomial families listed in Table 1 have been shown by Xiu [16] to provide convenient types of chaos based on their weight functions.

In general, the elements of the coordinate ($[\epsilon]$) are called the polynomial chaos coefficients. These coefficients hold deterministic information about the distribution of the random variable; for instance, the first and second central moments of ε can be calculated easily as

$$E[\varepsilon] = \mu_1 = \epsilon_0 \tag{10a}$$

$$E\left[(\varepsilon - E[\varepsilon])^2\right] = \mu_2 = \sum_{k=1}^{\infty} \epsilon_k^2 \langle \Psi_k^2 \rangle_{p(\xi)}, \tag{10b}$$

where $E[\ ]$ denotes expected value.

Now, let ε be an n-dimensional vector. Each of the n elements in ε are expanded separately; therefore, Equation (9) is effectively identical in vector form

$$\varepsilon(x,\xi) = \begin{bmatrix} \varepsilon^{(1)}(x^{(1)},\xi) \\ \varepsilon^{(2)}(x^{(2)},\xi) \\ \vdots \\ \varepsilon^{(n)}(x^{(n)},\xi) \end{bmatrix} = \begin{bmatrix} \sum_{k=0}^{\infty} \epsilon_k^{(1)}(x^{(1)})\Psi_k(\xi) \\ \sum_{k=0}^{\infty} \epsilon_k^{(2)}(x^{(2)})\Psi_k(\xi) \\ \vdots \\ \sum_{k=0}^{\infty} \epsilon_k^{(n)}(x^{(n)})\Psi_k(\xi) \end{bmatrix}.$$

Because the central moments are independent, the mean and variance of each their calculations similarly do not change. In addition to mean and variance, the correlation between two random variables is commonly desired. With the chaos coefficients estimated for each random variable and the polynomial basis known, correlation terms such covariance can be estimated.

3.1. Covariance

Let the continuous variables a and b have chaos expansions

$$a(x,\xi) = \sum_{j=0}^{\infty} \alpha_j(x)\Psi_j(\xi) \quad \text{and} \quad b(z,\zeta) = \sum_{k=0}^{\infty} \beta_k(z)\Phi_k(\zeta). \tag{11}$$

The covariance between a and b can be expressed in terms of two nested expected values

$$\text{cov}(a,b) = E[(a - E[a])(b - E[b])],$$

the external of which can be expressed as a double integral yielding

$$\text{cov}(a,b) = \int_{\mathbb{A}} \int_{\mathbb{B}} (a - E[a])(b - E[b])\, db\, da, \tag{12}$$

where $\mathbb{A}$ and $\mathbb{B}$ are the supports of a and b respectively. Substituting the expansions from Equation (11) into Equation (12) and acknowledging that the zeroth coefficient is the expected value gives

$$\text{cov}(a,b) = \int_{\mathbb{A}} \int_{\mathbb{B}} (ab - aE[b] - bE[a] + E[a]E[b])\, db\, da$$

$$= \int_{\mathbb{A}} \int_{\mathbb{B}} (ab - \beta_0 a - \alpha_0 b + \alpha_0 \beta_0)\, db\, da$$

$$= -\alpha_0 \beta_0 + \int_{\mathbb{A}} \int_{\mathbb{B}} (ab)\, db\, da \tag{13a}$$

$$= -\alpha_0 \beta_0 + \int_{\mathbb{X}} \int_{\mathbb{Z}} \sum_{j=0}^{\infty} \alpha_j(x)\Psi_j(\xi) \sum_{k=0}^{\infty} \beta_k(z)\Phi_k(\zeta)\, d\zeta\, d\xi. \tag{13b}$$

Note the change of variables between Equations (13a) and (13b). This is possible because the random variable and the weight function (a/ξ and b/ζ in this case) are over the same support. Additionally, the notation of the support variable is changed to be consistent with the integration variable.

As long as the covariance is finite, the summation and the integrals can be interchanged [49], giving a final generalized expression for the covariance to be

$$\operatorname{cov}(a,b) = \sum_{j=1}^{\infty} \sum_{k=1}^{\infty} \alpha_j(x)\beta_k(z) \int_{\mathbb{X}} \int_{\mathbb{Z}} \Psi_j(\xi)\Phi_k(\zeta) d\zeta d\xi. \tag{14}$$

In general, no further simplifications can be made; however, if the variables x and z are expanded using the same set of basis polynomials, then integration reduces to

$$\operatorname{cov}(a,b) = \sum_{k=1}^{\infty} \sum_{j=1}^{\infty} \alpha_k(x)\beta_j(z) \int_{\mathbb{X}} \Psi_k(\xi)\Psi_j(\xi) p(\xi) d\xi, \tag{15}$$

containing a single variable with respect to the base pdf. Taking advantage of the basis polynomial orthogonality yields the following simple expression:

$$\operatorname{cov}(a,b) = \sum_{k=1}^{\infty} \alpha_k(x)\beta_k(z) \langle \Psi_k^2 \rangle_{p(\xi)}. \tag{16}$$

Combined with the variance, the covariance matrix of the 2×2 system of x and z just discussed is given as

$$P = \sum_{k=1}^{\infty} \begin{bmatrix} \alpha_k^2 & \alpha_k \beta_k \\ \alpha_k \beta_k & \beta_k^2 \end{bmatrix} \langle \Psi_k^2 \rangle_{p(\xi)}.$$

For an n-dimensional state, let ϵ be the $n \times \infty$ matrix for the n, theoretically infinite, chaos coefficients. Written generally, the covariance matrix in terms of a chaos expansion is

$$P = \sum_{k=1}^{\infty} \epsilon_k \epsilon_k^T \langle \Psi_k^2 \rangle_{p(\xi)}.$$

In cases where orthonormal polynomials are used, the polynomial inner product disappears completely leaving only the summation of the estimated chaos coefficients

$$P = \sum_{k=1}^{\infty} \epsilon_k \epsilon_k^T. \tag{17}$$

3.2. Coefficient Calculation

The two most common methods of solving Equation (9) for the chaos coefficients are sampling-based and projection-based. The first, and most common, approach requires truncating the infinite summation in Equation (9) to yield

$$\varepsilon(x,\xi) = \sum_{k=0}^{N} \epsilon_k(x)\Psi_k(\xi), \tag{18}$$

where the truncation term N, which depends on the dimension of the state n and the highest order polynomial p, is

$$N+1 = \frac{(n+p)!}{n!p!}.$$

Drawing Q samples of ξ, where $Q > N$, and evaluating Ψ_k and ε at these points effectively results in randomly sampling ε directly. After initial sampling, ε can be transformed in x (commonly x is taken to be time so this indicates propagating the variable forward in time) resulting in a system of Q equations with $N+1$ unknowns that describe the pdf of ε after the transformation that is given by

$$\varepsilon(x,\xi_1) = \epsilon_0(x)\Psi_0(\xi_1) + \epsilon_1(x)\Psi_1(\xi_1) + \cdots + \epsilon_N(x)\Psi_N(\xi_1)$$
$$\varepsilon(x,\xi_2) = \epsilon_0(x)\Psi_0(\xi_2) + \epsilon_1(x)\Psi_1(\xi_2) + \cdots + \epsilon_N(x)\Psi_N(\xi_2)$$
$$\vdots$$
$$\varepsilon(x,\xi_Q) = \epsilon_0(x)\Psi_0(\xi_Q) + \epsilon_1(x)\Psi_1(\xi_Q) + \cdots + \epsilon_N(x)\Psi_N(\xi_Q).$$

This overdetermined system can be solved for ϵ using a least-squares approximation. The coefficients can then be used to calculate convenient statistical data about ε (e.g., central and raw moments).

While the sampling-based method is more practical to apply, the projection based method is not dependent on sampling the underlying distribution. Projecting the pdf of ε onto the jth basis yields

$$\langle \varepsilon(x,\xi), \Psi_j(\xi) \rangle_{p(\xi)} = \left\langle \sum_{k=0}^{\infty} \epsilon_k(x)\Psi_k(\xi), \Psi_j(\xi) \right\rangle_{p(\xi)}.$$

The inner product is with respect to the variable ξ; therefore, the coefficient ϵ acts as a scalar. The inner product is linear in the first argument; therefore, the summation coefficients can be removed from the inner product without alteration, that is

$$\langle \varepsilon(x,\xi), \Psi_j(\xi) \rangle_{p(\xi)} = \sum_{k=0}^{\infty} \epsilon_k(x) \langle \Psi_k(\xi), \Psi_j(\xi) \rangle_{p(\xi)}. \tag{19}$$

In contrast, if the summation is an element of the second argument, the linearity condition still holds; however, the coefficients incur a complex conjugate. Recall the basis polynomials are generally chosen to be orthogonal, so the right-hand inner product of Equation (19) reduces to the scaled Kronecker delta, resulting in

$$\langle \varepsilon(x,\xi), \Psi_j(\xi) \rangle_{p(\xi)} = \sum_{k=0}^{\infty} \epsilon_k(x) \langle \Psi_k(\xi), \Psi_j(\xi) \rangle_{p(\xi)}$$
$$= \sum_{k=0}^{\infty} \epsilon_k(x) c\, \delta_{kj}.$$

This leaves only the jth term (with the constant $c = \langle \Psi_j^2(\xi) \rangle_{p(\xi)}$), and an equation that is easily solvable for ϵ_j is

$$\epsilon_j(x) = \frac{\langle \varepsilon(x,\xi), \Psi_j(\xi) \rangle_{p(\xi)}}{\langle \Psi_j^2(\xi) \rangle_{p(\xi)}} \tag{20a}$$

$$= \frac{\int_{\mathbb{Z}} \varepsilon(x,\xi)\Psi_j(\xi) dp(\xi)}{\int_{\mathbb{Z}} \Psi_j^2(\xi) dp(\xi)}, \tag{20b}$$

which almost always requires numeric approximation.

3.3. Implementation Procedure

For convenience, the procedure for estimating the mean and covariance of a random state is given in Algorithm 1. Let ε be the state of a system with uncertainty defined by mean m and covariance P

subject to a set of system dynamics over the time vector T. The algorithm outlines the steps required to estimate the mean and covariance of the state after the amount of time specified by T.

Algorithm 1 Estimation of mean and covariance using a polynomial chaos expansion.

1: **procedure** PCE_EST(m_0, P_0)   ▷ Estimation of m and P using PCE
2:   **for** $k = 1$ **to** T **do**
3:    Draw samples of ξ based on chaos type   ▷ Either randomly or intelligently
4:    $\varepsilon_{k-1} \leftarrow m_{k-1}, P_{k-1}, \xi$ ▷ Sample ε based on ξ
5:    ε_k from propagating ε based on state dynamics
6:    $\epsilon \leftarrow$ Equation (20a)
7:    $m_k \leftarrow$ Equation (10a)
8:    $P_k \leftarrow$ Equation (17)
9:   **end for**
10:   **return** m, P
11: **end procedure**

3.4. Complex Polynomial Chaos

While polynomial chaos has been well-studied and applied to a various number of applications in $\mathbb{R}^n$, alterations must be made for the restricted space $\mathbb{S}^n$ due to its circular nature. A linear approximation can be made with little error when a circular variable's uncertainty is small; however, as the uncertainty increases, the linearization can impose significant error. Figure 2 shows the effects of projecting two wrapped normal distributions with drastically different standard deviations onto a tangent plane. The two wrapped normal distributions are shown in Figure 2a,b, with USTDs of 0.25 and 0.4 rad, respectively. Clearly, even relatively small USTDs result in approximately uniform wrapped pdfs.

One of the most basic projections is an orthogonal projection from an n-dimensional space onto an $n - 1$ dimensional plane. In this case, the wrapped normal pdf is projected orthogonally onto the plane $(1, x, z)$, which lies tangent to the unit circle at the point (1,0), coinciding with the mean direction of both pdfs. The plane, and the projection of the pdf onto this plane are shown in Figure 2c,d. Approximating the circular pdf as the projected planar pdf comes with an associated loss of information. At the tangent point, there is obviously no information loss; however, when the physical distance from the original point to the projected point is considered, the error associated with the projected point increases. As is the case with many projection methods concerning circular and spherical bodies, all none of the information from the far side of the body is available in the projection. The darkness of the shading in all of Figure 2 comes from the distance of the projection where white is no distance, and black is a distance value of least one (implying the location is on the hemisphere directly opposite the mean direction).

To better indicate the error induced by this type of projection, Figure 2e,f also include a measure that shows how far the pdf has been projected as a percentage of the overall probability at a given point. At the tangent point, there is no projection required, therefore the circular pdf has to be shifted 0% in the x direction. As the pdf curves away from the tangent plane, the pdf has to be projected farther. The difference between Figure 2e and Figure 2f is that the probability approaches zero nearing $y = \pm 1$ in Figure 2e; therefore, the effect of the error due to projection is minimal. In cases where the information is closely concentrated about one point, tangent plane projections can be good assumptions. Contrarily, in Figure 2f the pdf does not approach zero, and therefore the approximation begins to become invalid. Accordingly, the red error line approaches the actual pdf, indicating that the majority of the pdf has been significantly altered in the projection.

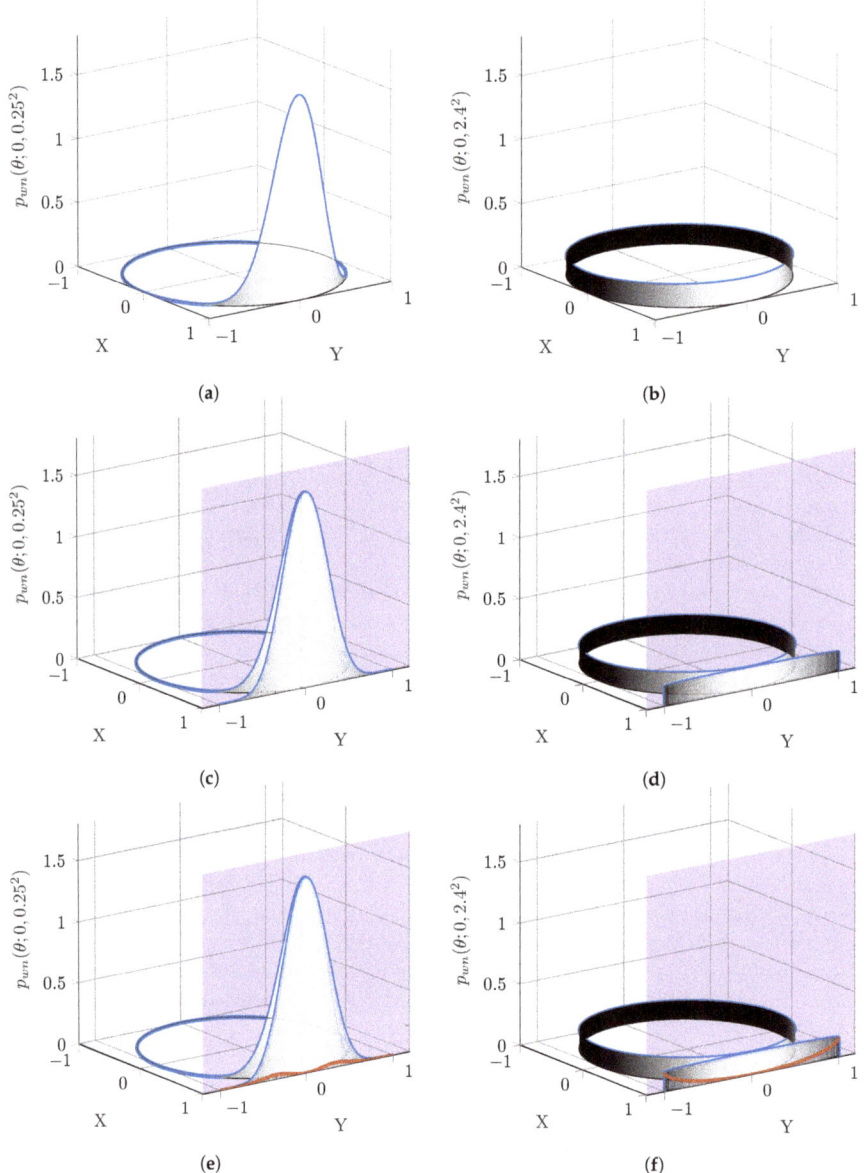

Figure 2. Error induced by approximating a circular distribution as linear, tangent at the mean. (**a**) Wrapped normal distribution with pdf $p_{wn}(\theta; 0, 0.25^2)$. (**b**) Wrapped normal distribution approaching wrapped uniform distribution with pdf $p_{wn}(\theta; 0, 2.4^2)$. (**c**) Projection of $p_{wn}(\theta; 0, 0.25^2)$ onto a plane tangent to the unit circle. (**d**) Projection of $p_{wn}(\theta; 0, 2.4^2)$ onto a plane tangent to the unit circle. (**e**) Error associated with projecting $p_{wn}(\theta; 0, 0.25^2)$ onto a plane tangent to the unit circle. (**f**) Error associated with projecting $p_{wn}(\theta; 0, 2.4^2)$ onto a plane tangent to the unit circle.

In addition to restricting the space to the unit circle, most calculations required when dealing with angles take place in the complex field. In truth, the bulk of expanding polynomial chaos to be suitable for angular random variables is generalizing it to complex vector spaces. Previous work by the authors [27] has shown that a stochastic angular random variable can be expressed using a polynomial chaos expansion. Specifically, the chaos expansion is one that uses polynomials that are orthogonal with respect to probability measures on the complex unit circle as opposed to the real line.

3.5. Szegő-Chaos

For the complex angular case, the chaos expansion is transformed slightly, such that

$$\varepsilon(x,\vartheta) = \sum_{k=0}^{\infty} \epsilon_k(x)\overline{\Psi_k(\vartheta)}, \tag{21}$$

where, once again, $\vartheta = e^{i\theta}$. The complex conjugate is not required in Equation (21), but it must be remembered that the expansion must be projected onto the *conjugate* of the expansion basis in Equation (20b). While ultimately a matter of choice, it is more convenient to express the expansion in terms of the conjugate basis, rather than the original basis.

Unfortunately, while the first moment is calculated the same for real and complex valued polynomials, the real valued process does not extend to complex valued polynomials. This is because of the slightly different orthogonality condition between real and complex valued polynomials. While the inner product given in Equation (2a) is not incorrect, it is only valid for real valued polynomials. The true inner product of two functions contains a complex conjugate, that is

$$\langle \Psi_m, \Psi_n \rangle_{p(x)} = \int_{\mathbb{X}} \Psi_m(x)\overline{\Psi_n(x)}p(x)dx = c\,\delta_{mn}.$$

The difference between $\mathbb{R}[x]$ and $\mathbb{C}[x]$ is that the complex conjugate has no effect on $\mathbb{R}[x]$. Fortunately, the zeroth polynomial of the Szegő polynomials is unitary just like the Askey polynomials. The complex conjugate has no effect; therefore the zeroth polynomial has no imaginary component and is calculated the same for complex and purely real valued random variables.

The complex conjugate of a real valued function has no effect; therefore, the first moment takes the form;

$$\mu_1 = \sum_{k=0}^{\infty} \epsilon_k(x) \int_{\mathbb{X}} \Psi_k(\xi)\Psi_0(\xi)p(\xi)d\xi = \sum_{k=0}^{\infty} \epsilon_k(x) \int_{\mathbb{X}} \Psi_k(\xi)\overline{\Psi_0(\xi)}p(\xi)d\xi. \tag{22}$$

In general, calculation of the second raw moment and the covariance cannot be simplified beyond

$$\mu_2 = \sum_{j=0}^{\infty}\sum_{k=0}^{\infty} \epsilon_j \epsilon_k \int_{\mathbb{X}} \overline{\Psi_j(\xi)\Psi_k(\xi)}p(\xi)d\xi$$

$$= \sum_{j=0}^{\infty}\sum_{k=0}^{\infty} \epsilon_j \epsilon_k \left\langle \overline{\Psi_j(\xi)}, \Psi_k(\xi) \right\rangle_{p(\xi)} \tag{23}$$

$$\text{cov}(x,z) = \sum_{j=1}^{\infty}\sum_{k=1}^{\infty} \alpha_j \beta_k \int_{\mathbb{X}}\int_{\mathbb{Z}} \overline{\Psi_j(\zeta)\Phi_k(\xi)}p(\zeta,\xi)d\zeta d\xi. \tag{24}$$

The simplification from Equation (14) to Equation (15) as a result of shared bases can similarly be applied to Equation (24). This simplifies Equation (24) to a double summation but only a singular inner product (i.e., integral), i.e.,

$$\text{cov}(x,z) = \sum_{j=1}^{\infty}\sum_{k=1}^{\infty} \alpha_j \beta_k \int_{\mathbb{X}} \overline{\Psi_j(\xi)\Psi_k(\xi)}p(\xi)d\xi$$

$$= \sum_{j=1}^{\infty}\sum_{k=1}^{\infty} \alpha_j \beta_k \langle \overline{\Psi_j(\xi)}, \Psi_k(\xi) \rangle_{p(\xi)}.$$

The familiar expressions for the second raw moment given in Equation (10b) and the covariance given in Equation (16) are special cases for $\mathbb{R}[x]$ rather than general expressions.

3.6. Rogers-Szegő-Chaos

The Rogers-Szegő polynomials and the wrapped normal distribution provide a convenient basis and random variable pairing for the linear combination in Equation (21). The Rogers-Szegő polynomials in Equation (3) can be rewritten according to [39]

$$\phi_n\left(-\frac{\vartheta}{\sqrt{q}}, q\right) = \sum_{k=0}^{n} (-1)^{n-k} \binom{n}{k}_q q^{\frac{n-k}{2}} \vartheta^k, \tag{25}$$

where q is calculated based on the standard deviation of the unwrapped normal distribution: $q = e^{-\sigma^2}$. These polynomials satisfy the orthogonality condition

$$\frac{1}{2\pi} \int_{-\pi}^{\pi} \phi_m\left(-\frac{\vartheta}{\sqrt{q}}, q\right) \overline{\phi_n\left(-\frac{\vartheta}{\sqrt{q}}, q\right)} \vartheta_3\left(\frac{\theta}{2}, \sqrt{q}\right) d\theta = (q; q)_n \delta_{mn},$$

where $\vartheta_3(\alpha, \beta)$ is the theta function

$$\vartheta_3(\alpha, \beta) = \sum_{k=-\infty}^{\infty} \beta^{k^2} e^{2ik\alpha}, \tag{26}$$

which is another form of the wrapped normal distribution. Note the distinction between the variables ϑ_3 and $\vartheta = e$.

For convenience, the inverse to the given theta function is

$$\vartheta_3^{-1}(\alpha, \beta) = 2\alpha + \pi + 2 \sum_{k=1}^{\infty} \frac{\beta^{k^2} \sin(2k\alpha)}{k}.$$

The inverse of the theta function is particularly useful if the cumulative distribution function (cdf) is required to draw random samples. The number of wrappings in Equation (26) significantly affects the results. For reference, the results presented in this work truncate the summation to ± 1000.

Written out, the first five orders of this form of the Rogers-Szegő polynomials are

$$\phi_0 = 1$$
$$\phi_1 = \vartheta - q^{1/2}$$
$$\phi_2 = \vartheta^2 - q^{1/2}(q+1)\vartheta + q$$
$$\phi_3 = \vartheta^3 - q^{1/2}(q^2+q+1)\vartheta^2 + q(q^2+q+1)\vartheta - q^{3/2}$$
$$\phi_4 = \vartheta^4 - q^{1/2}(q+1)(q^2+1)\vartheta^3 + q(q^2+1)(q^2+q+1)\vartheta^2 - q^{3/2}(q+1)(q^2+1)\vartheta + q^2,$$

and are shown graphically in Figure 3. Because the polynomials are complex valued, the real and imaginary components are show, separately. In both cases, the polynomials are oscillatory, with the real component being symmetric about $\theta = 0$, and the imaginary component being antisymmetric about $\theta = 0$. Additionally, the amplitude of the oscillations increase both with increasing order and distance from $\theta = 0$.

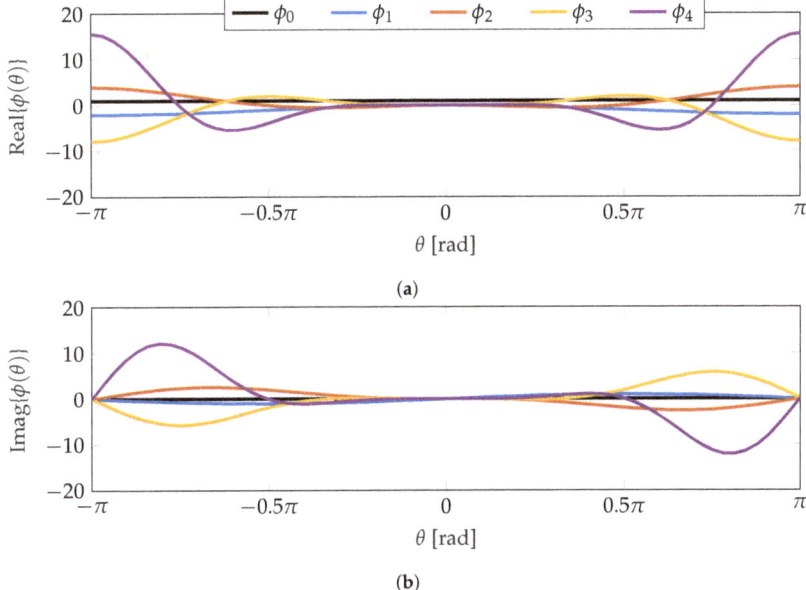

Figure 3. The zeroth through fourth Rogers-Szegő polynomials with an unwrapped standard deviation of 0.1. (**a**) Real Component. (**b**) Imaginary Component.

The zeroth polynomial is one, as is standard; therefore, the difference between the two generating functions given in Equations (25) and (26) will only be apparent in the calculation of moments beyond the first.

3.7. Function Complexity

As is to be expected, the computational complexity increases with increasing state dimension. It is therefore of interest to develop an expression that bounds the required number of function evaluations as a function of number of states and expansion order. Due to the many different methods of calculating inner products, all with different computational requirements, the number of functional inner products is what will be enumerated.

Let $x \in \mathbb{S}^P$ be a P-dimensional state vector consisting of angular variables, and let $q \in \mathbb{N}^P$ be the expansion order of each element in x, where $\mathbb{N}$ is the set of natural numbers, including zero. The number of inner products required to calculate the chaos coefficients in Equation (20b) for element x_i is $2(q_i + 1)$, where $\{i \in \mathbb{N} : k \leq P\}$ and q_i is the ith element of q.

Assume that the mean, variance, and covariance are desired for/between each element. The mean does not require any extra inner products, since the mean is simply the zeroth coefficient. The variance from Equation (23) requires an additional $(q_i + 1)^2$ inner products for a raw moment, or q_i^2 inner products for a central moment. Similarly, the covariance from Equation (24) between the ith and j^{th} elements requires $(q_i + 1)(q_j + 1)$ additional evaluations for a raw moment and $q_i q_j$ for a central moment. Combining these into one expression, the generalized number of inner product evaluations for raw moments with $P \geq 2$ is

$$2(q_1 + 1) + (q_1 + 1)^2 + \sum_{i=2}^{P} \left(2(q_i + 1) + (q_i + 1)^2 + \sum_{j=1}^{i-1} (q_i + 1)(q_j + 1) \right),$$

and for central moments is

$$2(q_1+1)+q_1^2+\sum_{i=2}^{P}\left(2(q_i+1)+q_i^2+\sum_{j=1}^{i-1}q_iq_j\right).$$

It should be noted that this is the absolute maximum number of evaluations that is required for an entirely angular state. In many cases inner products can be precomputed, the use of orthonormal polynomials reduces the coefficient calculation inner products by two, and expansions using real valued polynomials do not require these inner product calculations at all.

4. Numerical Verification and Discussion

To test the estimation methods outlined in Section 3.5, a system with two angular degrees of freedom is considered. The correlated, nonlinear dynamics governing this body, the initial mean directions ϕ/θ, initial USTDs, and constant angular velocities $\dot{\phi}/\dot{\theta}$ are given in Table 2.

Table 2. Initial conditions and governing equations of the dynamical system in Section 4.

Angle	Mean Direction	USTD	Angular Velocity (const.)	Dynamics
ϕ	$\pi/30$ [rad]	0.1 [rad]	0.2 [rad/s]	$\Delta\phi = \sqrt{|\theta_{k-1}|}\dot{\phi}\Delta T$
θ	$\pi/4$ [rad]	0.1 [rad]	0.2 [rad/s]	$\Delta\theta = \theta_{k-1}\dot{\theta}\Delta T$

For every simulation, the run time is 4 s with ΔT being 0.05 s; this equates to 81 time steps in each simulation. In Figure 4, the joint pdf propagates from the initial conditions (bottom left) to the final state (top right). The initial joint pdf clearly reflects an uncorrelated bivariate wrapped normal distribution. After being transformed by the dynamics, the final joint pdf exhibits not just translation and scaling, but also rotation: indicating a non-zero correlation between the two angles, which is desired.

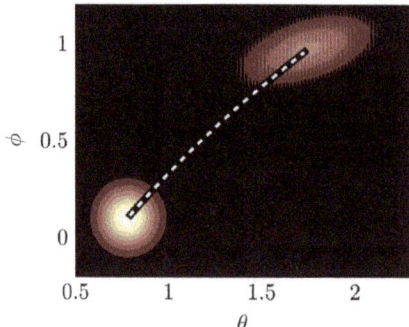

Figure 4. Evolution of the joint pdf from uncorrelated to correlated bivarite wrapped normal pdf.

For a practical application, the mean and standard deviation/variance of each dimension, as well as the covariance between dimensions is desired. When dealing with angles, the mean direction and the standard deviation can be obtained from the first moment, omitting the second moment. Therefore, only the first moment and the covariance will be discussed. Recall the equations for the first moment and covariance are in terms of chaos coefficients and are given generally in Equations (22) and (24). Because two angles are being estimated, the supports of the integrals in Equations (22) and (24) are set as $[-\pi, \pi)$: but it should be noted that the support is not rigidly defined this way, the only requirement is that the support length be 2π.

Rather than exploit the computational efficiency of methods such as quadrature integral approximations on the unit circle [50–52], the integrals are computed as Riemann sums. Therefore, it is necessary to determine an appropriate number of elements that provides an adequate numerical approximation, while remaining computationally feasible.

Figure 5 show the settling effect that decreasing the elemental angle has on the estimation of the covariance. Note that this figure is used to show the sensitivity of the simulation to the integration variable rather than the actual estimation of the covariance, which will be discussed later in this section. Both plots show the relative error of each estimate with respect to a Monte Carlo simulation of the joint system described in Table 2. Clearly, as the number of elements increases, the estimates begin to converge until the difference between $d\theta = 0.01$ rad (629 elements) and $d\theta = 0.005$ rad (1257 elements) is perceivable only near the beginning of the simulation. Because of this, it can reasonably be assumed that any error in the polynomial chaos estimate with $d\theta = 0.005$ is not attributed to numerical estimation of the integrals in Equations (22) and (24). Additionally, these results should also indicate the sensitivity of the estimate to the integration variable. Even though the dynamics used in this work's examples result in a joint pdf that somewhat resembles wrapped normal pdfs, the number of elements used in the integration must still be quite large. The final numerical element that must be covered is the Monte Carlo. For these examples, 5×10^7 samples are used in each dimension.

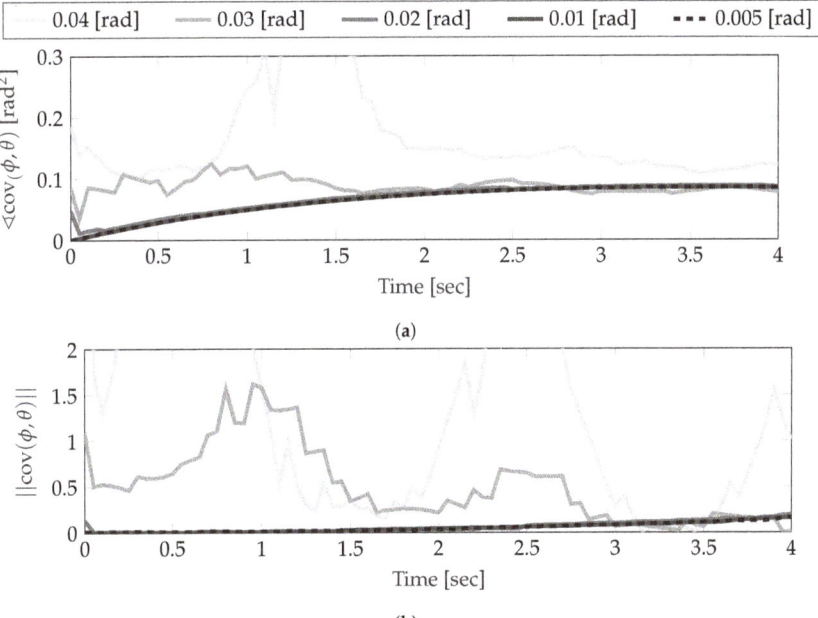

Figure 5. Effects of integration variable on estimated covariance relative error. Each line represents a different elemental angle ranging from 0.04 to 0.005 rad. (**a**) Angle Component. (**b**) Length Component.

In each of the examples in Section 4.1, the polynomial chaos estimate first evaluates the Rogers-Szegő polynomials at each of the 1257 uniformly distributed angles (ξ), solves Equation (20b) for the chaos coefficients, and uses Equations (22) and (24) to estimate the mean and covariance. After this, the 1257 realizations of the state ($\varepsilon(\xi)$) are propagated forward in time according to the system dynamics. At each time step the system is expanded using polynomial chaos to estimate the statistics.

4.1. Simulation Results

The estimations of the first moment and covariance of the system described by the simulation parameters in Table 2 are shown in Figures 6 and 7. In both cases, the angle and length of the estimate are presented, rather than the exponential form used in the polynomial chaos expansion. This representation much more effectively expresses the mean and concentration components of the estimate, which are directly of interest when examining statistical moments.

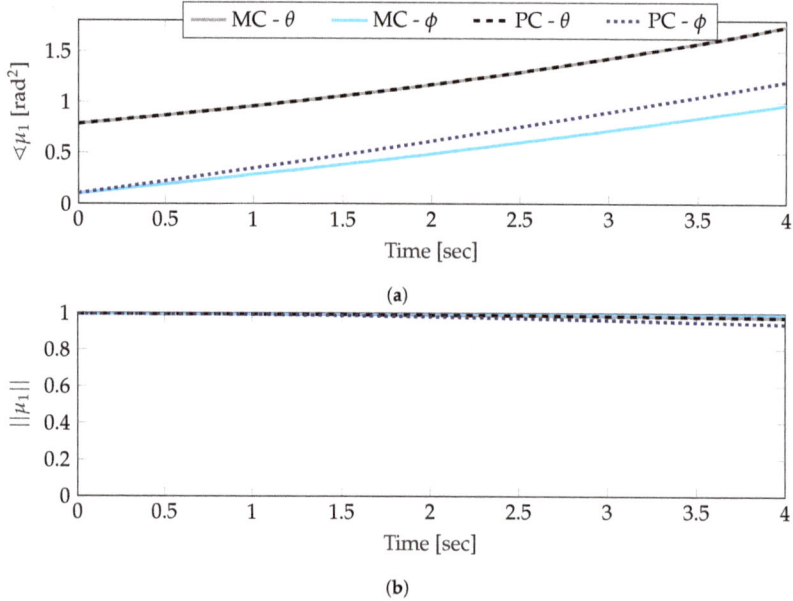

Figure 6. Estimation of the first moment of both angles using a fifth order polynomial chaos expansion compared against Monte Carlo simulation. (**a**) Angle Component. (**b**) Length Component.

Beginning with the mean estimate from a fifth order polynomial chaos expansion in Figure 6, the mean direction of the angle θ is nearly identical to the Monte Carlo estimate, while the estimate of ϕ begins to drift slightly as the simulation progresses. Of the two angles, this makes the most since; recalling Table 2, only the dynamics governing ϕ are correlated with θ, the dynamics governing θ are only dependent on θ. In comparison, the estimates of the lengths are both much closer to the Monte Carlo result. Looking closely at the end of the simulation, it can be seen that, again, θ is practically identical, and there is some small drift in ϕ downwards, indicating that the estimate reflects a smaller concentration. Effectively, the estimation of the mean reflects some inaccuracy; however, this inaccuracy is partly reflected in the larger dispersion of the pdf.

Similarly to the mean, a small drift can be seen in the estimate of the covariance in Figure 7. In both cases the initial estimate is nearly identical to the Monte Carlo result; however, throughout the simulation a small amount of drift becomes noticeable. While this drift is undesirable, the general tracking of the polynomial chaos estimate to the Monte Carlo clearly shows that the correlation between two angles can be approximated using a polynomial chaos expansion.

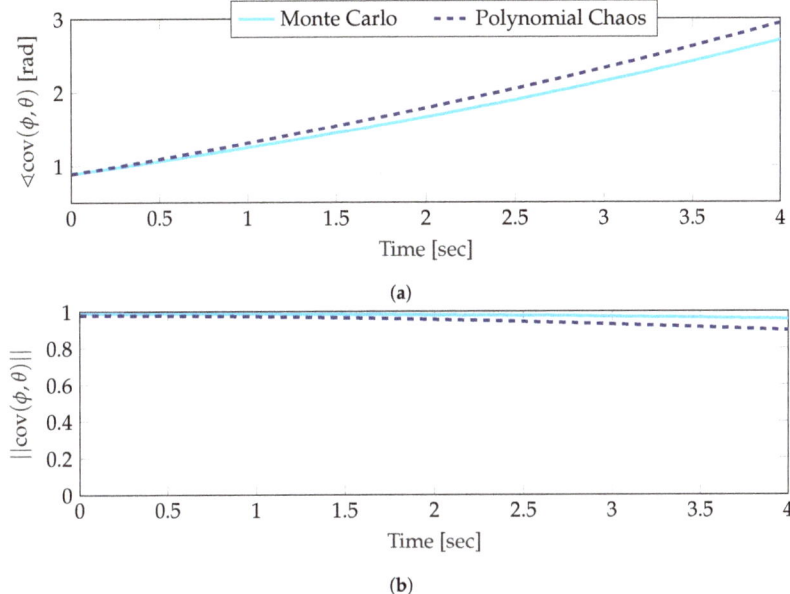

Figure 7. Estimation of the covariance between the two angles using a fifth order polynomial chaos expansion compared against Monte Carlo simulation. (**a**) Angle Component. (**b**) Length Component.

4.1.1. Unwrapped Standard Deviation and Joint PDF Assumptions

From the discussion of the generating function for the Rogers–Szegő polynomials Equation (25), it is clear that these polynomials are dependent on the USTD. Unfortunately this means that the polynomials are unique to any given problem, and while they can still be computed ahead of time and looked up, it is not as convenient as problems that use polynomials that are fixed (e.g., Hermite polynomials).

Additionally, the inner product in Equation (23), which describes the calculation of the covariance, requires the knowledge of the joint pdf between the two random variables. In practice, there is no reasonable way of obtaining this pdf; and if there is, then the two variables are already so well know, that costly estimation methods are irrelevant.

It is therefore of interest to investigate what effects making assumptions about the USTD and the joint pdf have on the estimates. The basis polynomials are evaluated when solving for the chaos coefficients Equation (20b) and when estimating the statistical moments Equations (22)–(25) at every time step. If no assumption is made about the USTD, then the generating function in Equation (25) or the three step recursion in Equation (5) must be evaluated at every time step as well. In either case, the computational burden can be greatly reduced if the basis polynomials remain fixed, requiring only an initial evaluation. Additionally, if the same USTD is used for both variables, than the simplification from two to one integrals in Equation (25) can be made.

While only used in the estimation of the covariance, a simplification of the joint pdf will also significantly reduce computation and increase the feasibility of the problem. The most drastic of simplifications is to use the initial, uncorrelated joint pdf. Note that the pdf used in the inner product is mean centered at zero (even for Askey chaos schemes); therefore, the validity of the estimation will not be effected by any movement of the mean.

Assuming the USTD to be fixed at 0.1 for both random variables and the joint pdf to be stationary throughout the simulation led to estimates that are within machine precision of the unsimplified results in Figures 6 and 7. This is to be expected when analyzing Askey-chaos schemes (like Hermite-chaos) that are problem invariant. In instances where the USTD of the wrapped normal distribution is low enough that probabilities at $\pm\pi$ are approximately zero, the wrapped normal distribution is effectively a segment of the unwrapped normal distribution, because the probabilities beyond $\pm\pi$ are approximately equal to zero. However, in problems where the USTD increases, the wrapped normal distribution quickly approaches a wrapped uniform distribution, this makes the time-invariant USTD a poor assumption. While a stationary USTD assumption may not hold as well for large variations in USTD, highly correlated, or poorly modeled, dynamical systems, it shows that some assumptions and simplifications can be made to ensure circular polynomial chaos is a practical method of estimation.

4.1.2. Chaos Coefficient Response

The individual chaos coefficients are not always inspected for problems using Askey-chaos simply due to the commonality of Askey-chaos. The adaptation of polynomial chaos to use the Szegő polynomials, and thus expanding from real to complex valued estimates presents a case that warrants inspection of the chaos coefficients. Figure 8 show the time evolution of the first 13 chaos coefficients (including the zeroth coefficient) that describe the random variable ϕ.

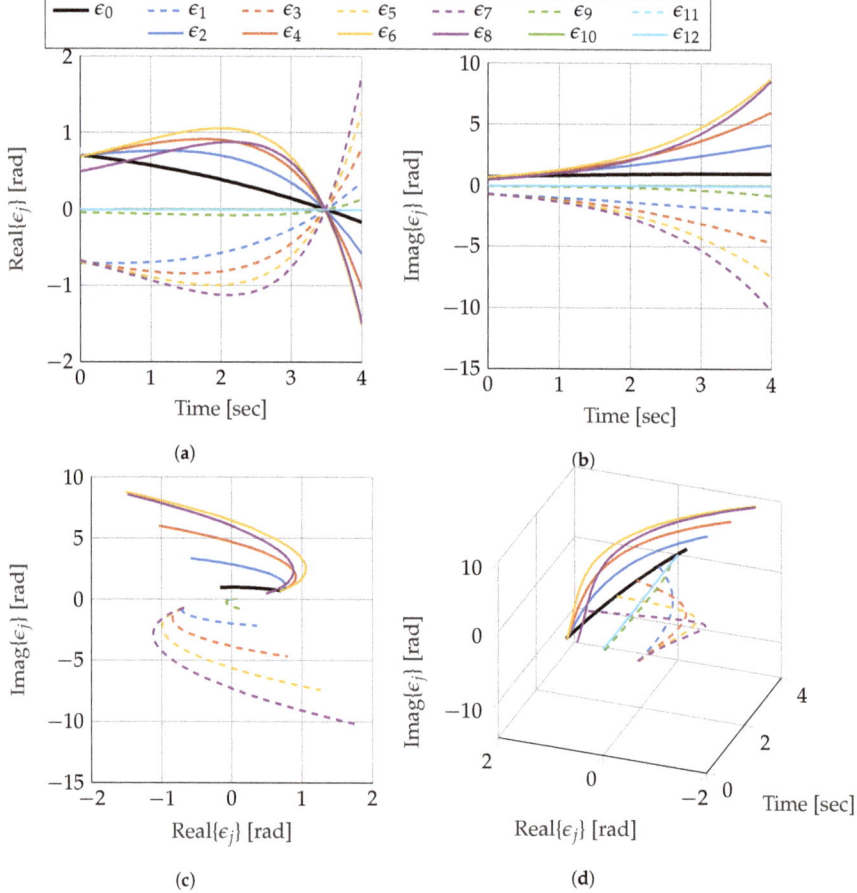

Figure 8. Time evolution of the first 13 chaos coefficients describing the random variable ϕ. (**a**) Real Coefficient Evolution. (**b**) Imaginary Coefficient Evolution. (**c**) Imaginary and Real Coefficient Evolution. (**d**) Complex Coefficient Evolution.

What becomes immediately apparent is that the coefficients are roughly anti-symmetrically paired until the length of the coefficient begins to approach zero. In this specific case, the eighth coefficient in Figure 8 initiates this trend. This is the first coefficient that does not have an initial estimate coinciding with lower ordered coefficients. All coefficients following this one show very little response to the system. This is to be expected. Recall the calculation of the chaos coefficient includes the product of polynomial and pdf as well as a division by the self inner product of each polynomial order (i.e., $\langle \Psi_k^2(\zeta) \rangle_{p(\zeta)}$). The polynomial and pdf product have opposing behaviors when approaching $\pm \pi$ from 0. Whereas the polynomial oscillation amplitude increases, the tails of the pdf approach probability values of zero. This ensures the growth in the higher order polynomials is mitigated.

For brevity, only the coefficients from the variable ϕ are shown. These have a much more interesting response than θ due to the nature of the dynamics. The most notable part of the coefficients from θ is that none of the coefficients ever move beyond the complex unit circle, which from Figure 8c, is clearly not the case for ϕ. In fact, the coefficients describing θ stay close to the complex unit circle and just move clockwise about it. Similarly, the eighth and higher coefficient lengths begin collapse to zero rad. For this problem (and presumably most problems) almost all of the information is coming from the first two coefficients. Comparing the estimates using two, three, and ten coefficients yields the same results to within machine precision. This is not surprising when considering the inner products (Table 3) that are required to estimate the covariance; each of the inner products are effectively zero when compared with $\langle \phi_0, \phi_0 \rangle_{p_{wn}}$ and $\langle \phi_1, \phi_1 \rangle_{p_{wn}}$. While having to only compute two significant chaos coefficients makes computation easier, it also limits the amount of information that is used in the estimate; however, for simple problems such as this one, two significant coefficients are satisfactory.

Table 3. Rogers-Szegő Inner Products.

$\langle \psi_i, \psi_j \rangle_{p_{wn}}$	$j = 0$	$j = 1$	$j = 2$	$j = 3$
$i = 0$	1.00	−4.77e-16−3.15e-15i	−3.35e-15−2.00e-16i	−1.21e-16−1.65e-16i
$i = 1$	−4.77e-16+3.15e-15i	−0.01−1.94e-16i	1.94e-4+3.06e-16i	−5.71e-06−3.27e-17i
$i = 2$	−3.35e-15+2.00e-16i	1.94e-4−3.06e-16i	1.84e-4+4.61e-17i	−1.09e-05−3.70e-17i
$i = 3$	−1.21e-16+1.65e-16i	−5.71e-06+3.27e-17i	−1.09e-05+3.70e-17i	−4.53e-06−2.45e-16i

5. Conclusions and Future Work

One method of quantifying the uncertainty of a random variable is a polynomial chaos expansion. For variables that exist only on the real line, this type of expansion has been well studied. This work developed the alterations that must be made for a polynomial chaos expansion to be valid for random variables that exist on the unit circle, specifically the complex unit circle (where the y coordinate becomes imaginary).

Previous work has shown that polynomial chaos can be used with the Rogers-Szegő polynomials to estimate the raw moments of a random variable with a wrapped normal distribution. A generalized set of expressions for the mean and covariance of multi-dimensional systems for both real and complex systems has been presented that do not make the assumption that each variable has been expanded with the same set of basis polynomials. An example of two angular random variables—one with correlated dynamics, and one without—has been presented. The mean of each random variable as well as the covariance between them is estimated and compared with Monte Carlo estimates. In the case of the uncorrelated random variable, the mean estimates are highly accurate. For the correlated random variable, the estimate is found to slowly diverge from the Monte Carlo result. A similar small divergence is observed in the covariance estimate; however, the general trend is similar enough to indicate the error is not in the formulation of the complex polynomial chaos. Additionally, an approximation to the basis polynomials and time-varying joint probability density function (pdf) is made, without loss of accuracy in the estimate. From the estimates of the mean and covariance, it is clear that the Rogers-Szegő polynomials can be used as an effective basis for angular random variable estimation. However, for more complex problems, different polynomials should be considered, specifically polynomials with an appropriate number of non-negligible self inner products.

Author Contributions: Conceptualization, C.S.; Methodology, C.S.; Formal Analysis, C.S.; Writing—Original Draft Preparation C.S.; Writing—Review & Editing, K.J.D.; Supervision, K.J.D. All authors have read and agreed to the published version of the manuscript.

Funding: This research was funded by the Graduate Assistance in Areas of National Need fellowship.

Conflicts of Interest: The authors declare no conflict of interest.

Abbreviations

The following abbreviations are used in this manuscript:

CDF   Cumulative distribution function
PCE   Polynomial chaos expansion
PDF   Probability density function
USTD   Unwrapped standard deviation

References

1. Haberberger, S.J. An IMU-Based Spacecraft Navigation Architecture Using a Robust Multi-Sensor Fault Detection Scheme. Master's Thesis, Missouri University of Science and Technology, Rolla, MO, USA, 2016.
2. Galante, J.; Eepoel, J.V.; Strube, M.; Gill, N.; Gonzalez, M.; Hyslop, A.; Patrick, B. Pose Measurement Performance of the Argon Relative Navigation Sensor Suite in Simulated-Flight Conditions. In Proceedings of the AIAA Guidance, Navigation, and Control Conference, Minneapolis, MN, USA, 13–16 August 2012.
3. Latella, C.; Lorenzini, M.; Lazzaroni, M.; Romano, F.; Traversaro, S.; Akhras, M.A.; Pucci, D.; Nori, F. Towards Real-Time Whole-Body Human Dynamics Estimation Through Probabilistic Sensor Fusion Algorithms. *Auton. Robot.* **2019**, *43*, 1591–1603. [CrossRef]
4. Lubey, D.P.; Scheeres, D.J. Supplementing state and dynamics estimation with information from optimal control policies. In Proceedings of the 17th International Conference on Information Fusion (FUSION), Salamanca, Spain, 7–10 July 2014; pp. 1–7.
5. Imani, M.; Ghoreishi, S.F.; Braga-Neto, U.M. Bayesian Control of Large MDPs with Unknown Dynamics in Data-Poor Environments. In *Advances in Neural Information Processing Systems 31*; Curran Associates, Inc.: Red Hook, NY, USA, 2018; pp. 8146–8156.
6. Hughes, D.L. *Comparison of Three Thrust Calculation Methods Using In-Flight Thrust Data*; NASA TM-81360; NASA: Washington, DC, USA, 1981; Unpublished.
7. Julier, S.; Uhlmann, J.; Durrant-Whyte, H.F. A New Method for the Nonlinear Transformation of Means and Covariances in Filters and Estimators. *IEEE Trans. Autom. Control* **2000**, *45*, 477–482. [CrossRef]
8. Wu, C.C.; Bossaerts, P.; Knutson, B. The Affective Impact of Financial Skewness on Neural Activity and Choice. *PLoS ONE* **2011**, *6*, e16838. [CrossRef] [PubMed]
9. Anderson, T.; Mattson, C. Propagating Skewness and Kurtosis Through Engineering Models for Low-Cost, Meaningful, Nondeterministic Design. *J. Mech. Des.* **2012**, *134*, 100911. [CrossRef]
10. Wiener, N. The Homogeneous Chaos. *Am. J. Math.* **1938**, *60*, 897–936. [CrossRef]
11. Janson, S. *Gaussian Hilbert Spaces*; Cambridge Tracts in Mathematics; Cambridge University Press: New York, NY, USA, 1997; Chapter 1.2.
12. Alexanderian, A. Gaussian Hilbert Spaces and Homogeneous Chaos: From theory to applications. 2013; Unpublished.
13. Eldred, M. Recent Advances in Non-Intrusive Polynomial Chaos and Stochastic Collocation Methods for Uncertainty Analysis and Design. In Proceedings of the 50th AIAA/ASME/ASCE/AHS/ASC Structures, Structural Dynamics, and Materials Conference, Palm Springs, CA, USA, 4–7 May 2009; Volume 4, pp. 2078–2114.
14. Ng, L.; Eldred, M. Multifidelity Uncertainty Quantification Using Non-Intrusive Polynomial Chaos and Stochastic Collocation. In Proceedings of the 53rd AIAA/ASME/ASCE/AHS/ASC Structures, Structural Dynamics and Materials Conference, Honolulu, HI, USA, 23–26 April 2012; Volume 9, pp. 7669–7685.

15. Savin, E.; Faverjon, B. Higher-Order Moments of Generalized Polynomial Chaos Expansions for Intrusive and Non-Intrusive Uncertainty Quantification. In Proceedings of the 19th AIAA Non-Deterministic Approaches Conference, Grapevine, TX, USA, 9–13 January 2017; pp. 215–223.
16. Xiu, D.; Karniadakis, G.E. The Wiener-Askey Polynomial Chaos for Stochastic Differential Equations. *SIAM J. Sci. Comput.* **2003**, *24*, 619–644. [CrossRef]
17. Schmid, C.L.; DeMars, K.J. Minimum Divergence Filtering Using a Polynomial Chaos Expansion. In Proceedings of the AAS/AIAA Astrodynamics Specialist Conference, Stevenson, WA, USA, 20–24 August 2017.
18. Xiu, D.; Karniadakis, G.E. Modeling Uncertainty in Flow Simulations via Generalized Polynomial Chaos. *J. Comput. Phys.* **2003**, *187*, 137–167. [CrossRef]
19. Hosder, S.; Walters, R.W.; Perez, R. A Non-Intrusive Polynomial Chaos Method for Uncertainty Propagation in CFD Simulations. In Proceedings of the 44th AIAA Aerospace Sciences Meeting and Exhibit, Reno, NV, USA, 9–12 January 2006.
20. Hosder, S.; Walters, R.W.; Balch, M. Point-Collocation Non-Intrusive Polynomial Chaos Method for Stochastic Computational Fluid Dynamics. *AIAA J.* **2010**, *48*, 2721–2730. [CrossRef]
21. Hosder, S.; Walters, R.W. Non-Intrusive Polynomial Chaos Methods for Uncertainty Quantification in Fluid Dynamics. In Proceedings of the 48th AIAA Aerospace Sciences Meeting, Orlando, FL, USA, 4–7 January 2010; pp. 1580–1595.
22. Jones, B.A.; Doostan, A.; Born, G.H. Nonlinear Propagation of Orbit Uncertainty Using Non-Intrusive Polynomial Chaos. *J. Guid. Control. Dyn.* **2013**, *36*, 430–444. [CrossRef]
23. Vittaldev, V.; Russell, R.P.; Linares, R. Spacecraft Uncertainty Propagation Using Gaussian Mixture Models and Polynomial Chaos Expansions. *J. Guid. Control. Dyn.* **2016**, *39*, 2615–2626. [CrossRef]
24. Mardia, K.; Jupp, P. *Directional Statistics*; Wiley Series in Probability and Statistics; John Wiley & Sons, Inc.: Hoboken, NJ, USA, 2009.
25. Markley, F.L. Attitude Error Representations for Kalman Filtering. *J. Guid. Control. Dyn.* **2003**, *26*, 311–317. [CrossRef]
26. Darling, J.E. Bayesian Inference for Dynamic Pose Estimation Using Directional Statistics. Ph.D. Thesis, Missouri University of Science and Technology, Rolla, MO, USA, 2016.
27. Schmid, C.L.; DeMars, K.J. Polynomial Chaos Confined to the Unit Circle. In Proceedings of the AAS/AIAA Astrodynamics Specialist Conference, Snowbird, UT, USA, 19–23 August 2018; pp. 2239–2256.
28. Jones, B.A.; Balducci, M. Uncertainty Propagation of Equinoctial Elements Via Stochastic Expansions. In Proceedings of the John L. Junkins Dynamical Systems Symposium, College Station, TX, USA, 20–21 May 2018.
29. Andrews, G.E.; Askey, R. *Classical Orthogonal Polynomials. Polynômes Orthogonaux et Applications*; Brezinski, C., Draux, A., Magnus, A.P., Maroni, P., Ronveaux, A., Eds.; Springer: Berlin/Heidelberg, Germany, 1985; pp. 36–62.
30. Askey, R.; Wilson, J. *Some Basic Hypergeometric Orthogonal Polynomials that Generalize Jacobi Polynomials*; Number 319 in American Mathematical Society: Memoirs of the American Mathematical Society; American Mathematical Society: Providence, RI, USA, 1985; Chapter 1.
31. Koekoek, R.; Lesky, P.A.; Swarttouw, R.F. *Hypergeometric Orthogonal Polynomials and Their q-Analogues*; Springer Monographs in Mathematics; Springer: Berlin/Heidelberg, Germany, 2010.
32. Szegő, G. *Orthogonal Polynomials*; American Mathematical Society Colloquium Publications, American Mathematical Society: Providence, RI, USA, 1959; Volume 23, Chapter 11, pp. 287–295.
33. Simon, B. *Orthogonal Polynomials on the Unit Circle (Colloquium Publications)*; American Mathematical Society: Providence, RI, USA, 2005; Volume 54.
34. Simon, B. Orthogonal Polynomials on the Unit Circle: New Results. *arXiv* **2004**, arXiv:math.SP/math/0405111.
35. Geronimus, Y.L. Orthogonal Polynomials: Estimates, Asymptotic Formulas, and Series of Polynomials Orthogonal on the Unit Circle and on an Interval. *Math. Gaz.* **1962**, *46*, 354–355. [CrossRef]
36. Ismail, M.E.; Ruedemann, R.W. Relation Between Polynomials Orthogonal on the Unit Circle with Respect to Different Weights. *J. Approx. Theory* **1992**, *71*, 39–60. [CrossRef]
37. Jagels, C.; Reichel, L. On the Construction of Szegő Polynomials. *J. Comput. Appl. Math.* **1993**, *46*, 241–254. [CrossRef]

38. Jones, W.B.; Njåstad, O.; Thron, W.J. Moment Theory, Orthogonal Polynomials, Quadrature, and Continued Fractions Associated with the Unit Circle. *Bull. Lond. Math. Soc.* **1989**, *21*, 113–152. [CrossRef]
39. Atakishiyev, N.M.; Nagiyev, S.M. On the Rogers-Szegő Polynomials. *J. Phys. Math. Gen.* **1994**, *27*, L611. [CrossRef]
40. Hou, Q.; Lascoux, A.; Mu, Y. Continued Fractions for Rogers-Szegő Polynomials. *Numer. Algorithms* **2004**, *35*, 81–90. [CrossRef]
41. Vinroot, C.R. Multivariate Rogers-Szegő Polynomials and Flags in Finite Vector Spaces. *arXiv* **2010**, arXiv:math.CO/1011.0984.
42. Szabłowski, P.J. On q-Hermite Polynomials and Their Relationship With Some Other Families of Orthogonal Polynomials. *Demonstr. Math.* **2011**, *46*, 679–708. [CrossRef]
43. Andrews, G. *The Theory of Partitions*; Cambridge Mathematical Library, Cambridge University Press: Cambridge, UK, 1998.
44. Fisher, R. Dispersion on a Sphere. *Proc. R. Soc. London. Ser. Math. Phys. Eng. Sci.* **1953**, *217*, 295–305. [CrossRef]
45. Watson, G.S.; Williams, E.J. On the Construction of Significance Tests on the Circle and the Sphere. *Biometrika* **1956**, *43*, 344–352. [CrossRef]
46. Mardia, K.V. Distribution Theory for the von Mises-Fisher Distribution and Its Application. In *A Modern Course on Statistical Distributions in Scientific Work*; Patil, G.P., Kotz, S., Ord, J.K., Eds.; Springer: Dordrecht, The Netherlands, 1975; pp. 113–130.
47. Bogachev, V. *Gaussian Measures*; Mathematical Surveys and Monographs, American Mathematical Society: Providence, RI, USA, 2015.
48. Cameron, R.H.; Martin, W.T. The Orthogonal Development of Non-Linear Functionals in Series of Fourier-Hermite Functionals. *Ann. Math.* **1947**, *48*, 385–392. [CrossRef]
49. Fubini, G. *Sugli Integrali Multipli: Nota*; Reale Accademia dei Lincei: Rome, Italy, 1907.
50. Cruz-Barroso, R.; Mendoza, C.D.; Perdomo-Pío, F. Szegő-Type Quadrature Formulas. *J. Math. Anal. Appl.* **2017**, *455*, 592–605. [CrossRef]
51. Bultheel, A.; González-Vera, P.; Hendriksen, E.; Njåstad, O. Orthogonal Rational Functions and Quadrature on the Unit Circle. *Numer. Algorithms* **1992**, *3*, 105–116. [CrossRef]
52. Daruis, L.; González-Vera, P. Szegő Polynomials and Quadrature Formulas on the Unit Circle. *Appl. Numer. Math.* **2001**, *36*, 79–112.. [CrossRef]

© 2020 by the authors. Licensee MDPI, Basel, Switzerland. This article is an open access article distributed under the terms and conditions of the Creative Commons Attribution (CC BY) license (http://creativecommons.org/licenses/by/4.0/).

Article

Bivariate Thiele-Like Rational Interpolation Continued Fractions with Parameters Based on Virtual Points

Le Zou [1,2,3,*], Liangtu Song [2,3], Xiaofeng Wang [1,*], Yanping Chen [1], Chen Zhang [1] and Chao Tang [1]

1. School of Artificial Intelligence and Big Data, Hefei University, Hefei 230601, China; chenypo@hfuu.edu.cn (Y.C.); zhangchen@hfuu.edu.cn (C.Z.); tangchao@hfuu.edu.cn (C.T.)
2. Institute of Intelligent Machines, Hefei Institutes of Physical Science, Chinese Academy of Sciences, Hefei 230031, China; ltsong@iim.ac.cn
3. University of Science and Technology of China, Hefei 230026, China
* Correspondence: zoule@mail.ustc.edu.cn (L.Z.); xfwang@hfuu.edu.cn (X.W.)

Received: 24 December 2019; Accepted: 31 December 2019; Published: 2 January 2020

Abstract: The interpolation of Thiele-type continued fractions is thought of as the traditional rational interpolation and plays a significant role in numerical analysis and image interpolation. Different to the classical method, a novel type of bivariate Thiele-like rational interpolation continued fractions with parameters is proposed to efficiently address the interpolation problem. Firstly, the multiplicity of the points is adjusted strategically. Secondly, bivariate Thiele-like rational interpolation continued fractions with parameters is developed. We also discuss the interpolant algorithm, theorem, and dual interpolation of the proposed interpolation method. Many interpolation functions can be gained through adjusting the parameter, which is flexible and convenient. We also demonstrate that the novel interpolation function can deal with the interpolation problems that inverse differences do not exist or that there are unattainable points appearing in classical Thiele-type continued fractions interpolation. Through the selection of proper parameters, the value of the interpolation function can be changed at any point in the interpolant region under unaltered interpolant data. Numerical examples are given to show that the developed methods achieve state-of-the-art performance.

Keywords: Thiele-like rational interpolation continued fractions with parameters; unattainable point; inverse difference; virtual point

---

1. Introduction

The interpolation method plays a critical role in approximation theory and is a classical topic in numerical analysis [1–6]. It is believed that interpolation polynomial and rational interpolation are two popular interpolation methods. They have many applications, such as image interpolation processing, numerical approximation [1], extensive applications in terms of arc structuring [1,2], and so on. Kuchminska et al. [3] proposed a Newton-Thiele-like Interpolating formula for two variate interpolation. Pahirya et al. [4] developed the problem of the interpolant function of bivariate by two-dimensional continued fractions. Li et al. [5] generalized Thiele's expansion of a univariate rational interpolation function to the Thiele-Newton blending rational interpolation. The authors also developed the Viscovatov-like algorithm to calculate the coefficients of Thiele-Newton's expansion. Cuyt et al. [6] used a multivariate data fitting technique to solve various scientific computing problems in filtering, meta-modelling, queueing, computational finance, networks, graphics, and more. Li et al. [7] analyzed the fractional-order unified chaotic system by different fractional calculus numerical methods. Xu et al. [8] introduced the truncated exponential radial function for surface modeling. Massopust [9] constructed non-stationary fractal functions and interpolation based on non-stationary versions of fixed points. In recent years, different aspects of rational multivariate interpolation were studied, especially in Newton form. Cuyt et al. [10] define multivariate divided differences of the multivariate Newton–Padé approximants. Akal et al. [11] modified Cuyt and Verdonk's approach to multivariate Newton–Padé approximations. Ravi [12] studied the minimal rational interpolation problem using algebrogeometric methods. Bertrand et al. [13] proposed a new polynomial projector on a space of functions of many variables, and studied the main approximation properties of the new projectors. One multipoint multivariate polynomial interpolation method from the Goodman–Hakopian polynomial interpolation was generalized to the case of rational interpolation in the paper [14]. The authors presented the scale of mean value multipoint multivariate Padé interpolations which includes as particular cases both the scale of mean value polynomial interpolations and the multipoint multivariate Padé approximations. Based on the collocation polynomial and Hermite interpolation, Li et al. [15] proposed a numerical explicit gradient scheme, which has higher convergence order.

The interpolation method also was applied to graphic image morphing and image processing [16–24]. In the paper [16], an effective directional Bayer color filter array demosaicking method based on residual interpolation is presented. The proposed algorithm guaranteed the quality of color images and reduced the computational complexity. Zhou et al. [17] developed an interpolation filter called an all-phase discrete sine transform filter and used it for image demosaicking. Min et al. [18] proposed a nonlinear approximation method based on Taylor series to describe and approximate images with intensity inhomogeneity. He et al. [19,20] presented a Thiele-Newton's rational interpolation function in the polar coordinates which was then applied to image super-resolution. The new method had a lower time cost and a better magnified effect. Yao et al. [21] proposed a new approach to bivariate rational interpolation. The presented interpolation method was identified by the values of shape parameters and scaling factors from the paper [21]. Zhang et al. [22] presented a single-image super-resolution algorithm based on rational fractal interpolation. Based on a multi-scale optical flow reconstruction scheme, Ahn et al. [23] proposed a fast 4K video frame interpolation method. Wei et al. [24] adopted the bilinear interpolation to obtain the Region of Interest pooling layer in image manipulation detection. In recent years, Zhang et al. [25,26] have reported on some new types of weighted blending spline interpolation. By selecting different coefficients and appropriate parameters, the value of the spline interpolation function can be modified at any point, in the interpolant region under unchanging interpolant data, so the interpolation functions of geometric surfaces can be adjusted, even for the given data, in actual design, but the computation is complicated. One of the authors proposed an associated continued fractions rational interpolation and its Viscovatov algorithm through Taylor expansion, proposed some different types of bivariate interpolation, and studied several general interpolation functions [27]. Tang and Zou [28,29] studied and provided some general interpolation frames with many interpolation formulae. For the given interpolation data, it could handle some special interpolant problems through selecting parameters appropriately. However, it is still a quite difficult problem to select appropriate parameters and then design the interpolation format while meeting the conditions in the computational process. It is difficult to determine such a function without the process of comparison and trial. Zhao et al. [30] presented the block-based Thiele-like blending rational interpolation. For the given interpolation points, many types of block-based Thiele-like blending rational interpolation were studied based on different block points of data. For image processing and computer-aided geometric design, there is still substantial demand for complicated models and the integration of design and fabrication, but two problems remain:

1. How to construct a proper polynomial or rational interpolation with explicit mathematical expression and a simple calculation which make the function easy to use and convenient to study theoretically;
2. For the given data, how to modify the curve or surface shape to enable the function to meet the actual requirements.

In the classical interpolation method, the interpolation function is unique to the interpolation points, and it is almost impossible to resolve the above two problems.

Thus, this raises an interesting question: whether many unique rational interpolations based on inverse difference exist. At present, the Thiele-type rational interpolation continued fractions is the hot topic regarding methods of rational interpolation; however, it is unique for the given interpolation data, and this limits its application. The Thiele-like rational interpolation continued fractions may meet nonexistent inverse differences and unattainable points. To avoid the problems mentioned above, this paper aims to develop bivariate Thiele-like rational interpolation continued fractions by introducing one or more parameters, which can adjust the shape of the curves or surfaces without altering the given interpolation points, so as to meet the practical requirements. Meanwhile, in contrast to the classical interpolation method in [1], our method can avoid unattainable points and nonexistent inverse differences in interpolation problems and performs better. In contrast to the interpolation methods presented in [1,28,31,32], it is unnecessary to adjust the nodes, and only the addition of multiple numbers in the sight of unattainable points is required, which makes it simple to numerate. To solve the above problem, in the paper [33–35], the authors developed a univariate Thiele-like rational interpolation continued fractions with parameters. The question can be solved with the proposed method in the paper, but the authors only discussed the univariate case. We generalize the results to the bivariate case in this paper.

The organization of the paper is as follows: We give a brief review on univariate Thiele-like rational interpolation continued fractions with parameters and discuss a special rational interpolation problem where unattainable points and inverse differences do not exist, and solve it through univariate Thiele-like rational interpolation with parameters in Section 2. We propose four bivariate Thiele-like branched rational interpolation continued fractions with parameters in Section 3. In addition to the bivariate Thiele-like branched rational interpolation continued fractions with unattainable points, the dual interpolation format of bivariate Thiele-like branched rational interpolation continued fractions with a single parameter and bivariate Thiele-like branched rational interpolation continued fractions with a nonexistent partial inverse difference are also discussed. As an application of the proposed methods, numerical examples are given to illustrate the effectiveness of the methods in Section 4.

2. Univariate Thiele-Like Interpolation Continued Fractions with Parameters

Let us consider the following univariate rational interpolation problem. If we consider a function $y = f(x)$ and support points set $\{(x_0,y_0),(x_1,y_1),\ldots(x_n,y_n)\}$ on interval $[a,b]$, we can gain the following classical Thiele-like rational interpolation continued fractions [1]:

$$R_n(x) = b_0 + \frac{x - x_0}{b_1} + \frac{x - x_1}{b_2} + \cdots + \frac{x - x_{n-1}}{b_n} \tag{1}$$

where $b_i\, (i = 1,\ldots,n)$ represents the inverse differences of $f(x)$.

2.1. Thiele-Like Rational Interpolation Continued Fractions with Parameter

It is well known from the literature that if Thiele-type continued fractions rational interpolation functions exist, they are unique compared with the popular method. This is inconvenient for practical application. To solve this problem, many scholars have proposed several improved methods. Zhao et al. [25] demonstrated the block-based Thiele-like blending rational interpolation. For a given set of interpolation points, many kinds of Thiele-type continued fractions interpolation functions can be constructed based on different block points. However, since every interpolation function is constructed for a special block method, one cannot derive different interpolation functions for special block points, and the interpolation function cannot adjust and may meet unattainable points. For a special block-based method, one must construct a polynomial or rational interpolation and calculate the block-based inverse difference and then construct the block-based Thiele-like blending rational interpolation. This requires a large amount of calculation and is inconvenient for the interpolation application. So, Zou et al. [33–35] constructed several novel univariate Thiele-like rational interpolation continued fractions with parameters, which has many advantages. By introducing a new parameter, λ ($\lambda \neq 0$), the authors [33,34] considered taking a point of the original points (x_k, y_k) ($k = 0, 1, \ldots, n$) as a virtual double point, and the multiplicity of the other points remains the same.

Let
$$y_i^0 = y_i, \ i = 0, 1, \ldots, n. \tag{2}$$

Suppose $k < n$, when $j = 1, \ldots, k+1$, for $i = j, j+1, \ldots, n$,
$$y_i^j = \frac{x_i - x_{j-1}}{y_i^{j-1} - y_{j-1}^{j-1}}. \tag{3}$$

For $i = k+1, k+2, \ldots, n$,
$$z_i^{k+1} = \frac{x_i - x_k}{y_i^{k+1} - \frac{1}{\lambda}}. \tag{4}$$

When $j = k+2, k+3, \ldots, n$, for $i = j, j+1, \ldots, n$,
$$z_i^j = \frac{x_i - x_{j-1}}{z_i^{j-1} - z_{j-1}^{j-1}}. \tag{5}$$

The Thiele-like rational interpolation continued fractions with a single parameter have the formula as follows:

$$R_n^{(0)}(x) = c_0 + \frac{x - x_0}{c_1} + \cdots + \frac{x - x_{k-1}}{c_k} + \frac{x - x_k}{c_{k+1}^0} + \frac{x - x_k}{c_{k+1}} + \frac{x - x_{k+1}}{c_{k+2}} + \cdots + \frac{x - x_{n-1}}{c_n}, \tag{6}$$

where
$$c_i = \begin{cases} y_i^i, \ i = 0, 1, \ldots, k, \\ z_i^i, \ i = k+1, k+2, \ldots, n, \end{cases}$$

$$c_{k+1}^0 = \frac{1}{\lambda}. \tag{7}$$

Without loss of generality, the authors [33,34] generalized the results to the Thiele-like rational interpolation continued fractions with two parameters. The authors discussed two categories: an arbitrary point of the original points is considered as a treble virtual point; two arbitrary points of original points are considered as the virtual double points. It can be seen that the new kind of Thiele-like continued fraction is not unique, and it satisfies the given interpolation condition. We know that it could meet nonexistent inverse differences and unattainable points in the classical Thiele-type continued fractions interpolation. As a fact, the Thiele-like rational interpolation continued fractions with parameters can solve the above interpolation problem. We discuss this problem in the next two subsections.

2.2. The Interpolation Problem with Unattainable Points

Definition 1 ([31]). *Suppose the given point is $D = \{(x_i, y_i) | i = 0, 1, \ldots, n\}$ and x_i is diverse, $R(x) = \frac{N(x)}{D(x)}$ is the Thiele-type interpolation continued fractions in Formula (1) if (x_k, y_k) satisfies*
$$N(x_k) - D(x_k)y_k = 0, R(x_k) = \frac{N(x_k)}{D(x_k)} \neq y_k. \tag{8}$$

Then, (x_k, y_k) is an unattainable point of $R(x)$.

Theorem 1 ([31,32]). *Suppose the given point is $D = \{(x_i, y_i) | i = 0, 1, \ldots, n\}$, and x_i is diverse, the Thiele-type interpolation continued fraction is as shown in Formula (1), where $b_k \neq \infty$, $(k = 0, 1, \ldots, n-1)$, $b_n \neq 0$, and then the necessary and sufficient condition of (x_k, y_k) for an unattainable point is*

$$R_k(x_k) = 0, \tag{9}$$

where $R_k(x) = \frac{N_k(x)}{D_k(x)}$ is irreducible and

$$R_{n-1}(x) = b_n, \; R_i(x) = b_{i+1} + (x - x_{i+1}) R_{i+1}^{-1}(x), i = n-2, n-3, \ldots, k. \tag{10}$$

Theorem 2. *The Thiele-like rational interpolation continued fractions with a single parameter defined by Equations (2)–(5) satisfies*

$$R_n^{(0)}(x_k) = y_k. \tag{11}$$

Proof. From Equation (6), let $R_k^{(0)}(x) = \frac{1}{\lambda} + \frac{x - x_k}{R_{k+1}^{(0)}(x)}$, where

$$R_n^{(0)}(x) = c_n,$$

$$R_i^{(0)}(x) = c_i + \frac{x - x_i}{R_{i+1}^{(0)}(x)} \; (i = n-1, n-2, \ldots, k+1).$$

Then, $R_k^{(0)}(x_k) = \frac{1}{\lambda} \neq 0$ from Theorem 1, (x_k, y_k) is an unattainable point of $R_n^{(0)}(x)$, and then we have $R_n^{(0)}(x_k) = y_k$.

The proof is complete. □

2.3. The Thiele-Like Rational Interpolation Continued Fractions Problem with a Nonexistent Inverse Difference

Given diverse interpolation data $\{(x_0, y_0), (x_1, y_1), \ldots, (x_n, y_n)\}$, in the process of constructing Thiele-like rational interpolation continued fractions, the inverse difference would be ∞ if the denominator equals zero, i.e., the inverse difference does not exist, which results in the failure of Thiele-type continued fractions interpolation function. Considering this case, assume that y_{k+1}^{k+1} does not exist (i.e., $y_{k+1}^{k+1} = \infty$), we introduce a parameter η ($\eta \neq 0$), and construct the novel inverse difference as shown in Table 1.

From Equation (4), we can get

$$z_{k+1}^{k+1} = \frac{x_{k+1} - x_k}{y_{k+1}^{k+1} - \frac{1}{\eta}} = \frac{x_{k+1} - x_k}{\infty - \frac{1}{\eta}} = 0. \tag{12}$$

Using the method given in Formula (6), we can construct a Thiele-like rational interpolation continued fractions with a parameter:

$$R_n^{(3)}(x) = c_0 + \frac{x - x_0}{c_1} + \cdots + \frac{x - x_{k-1}}{c_k} + \frac{x - x_k}{c_{k+1}^0} + \frac{x - x_k}{c_{k+1}} + \frac{x - x_{k+1}}{c_{k+2}} + \cdots + \frac{x - x_{n-1}}{c_n}. \tag{13}$$

Additionally, the calculating method of c_i ($i = 0, 1, \ldots, n$) follows Formulas (2)–(5), $c_{k+1}^0 = \frac{1}{\eta}$.

It is easy to prove that $R_n^{(3)}(x)$ satisfies the interpolation condition.

For the special interpolation problems discussed in Sections 2.2 and 2.3, there are four methods to overcome them: (a) adjust the interpolation nodes [1,30]; (b) replace the inverse difference by divided differences [36]; and (c) replace the inverse difference by block-based inverse differences [28,36,37]. In addition, there is also a method provided through the selection parameter in papers [27–29].

Compared with the methods above, it is easy to see that the method in this paper is simpler and more convenient.

Table 1. Inverse differences table where an inverse difference does not exist.

Nodes	0 Order Inverse Differences	1 Order Inverse Differences	...	k	k + 1 Order Inverse Differences	k + 2 Order Inverse Differences	...	n + 1 Order Inverse Differences
x_0	y_0^0							
x_1	y_1^0	y_1^1						
$\vdots$	$\vdots$	$\vdots$	$\ddots$					
x_k	y_k^0	y_k^1	...	y_k^k				
x_k	y_k^0	y_k^1	...	y_k^k	$\frac{1}{\eta}$			
x_{k+1}	y_{k+1}^0	y_{k+1}^1	...	y_{k+1}^k	y_{k+1}^{k+1}	z_{k+1}^{k+1}		
$\vdots$	$\vdots$	$\vdots$	...	$\vdots$	$\vdots$	$\vdots$	$\ddots$	
x_n	y_n^0	y_n^1	...	y_n^k	y_n^{k+1}	z_n^{k+1}	...	z_n^n

3. Multivariate Thiele-Like Branched Rational Interpolation Continued Fractions with Parameters

Now, we generalize the previous methods to the computation of the multivariate case. For simplicity, and also without loss of generality, we restrict ourselves to the case where bivariate problems are involved.

Suppose $\prod_{m,n} \subset D \subset R^2$ is the diverse rectangular net on rectangular region D, $f(x,y)$ is the real function defined on rectangular region D, and let

$$f(x_i, y_j) = f_{i,j}, i = 0, 1, \ldots, m, j = 0, 1, \ldots, n. \tag{14}$$

The bivariate Thiele-like branched rational interpolation continued fractions is as follows:

$$R(x,y) = b_{0,0} + \frac{y-y_0}{b_{0,1}} + \frac{y-y_1}{b_{0,2}} + \cdots + \frac{y-y_{n-1}}{b_{0,n}} + \cfrac{x-x_0}{b_{1,0} + \frac{y-y_0}{b_{1,1}} + \frac{y-y_1}{b_{1,2}} + \cdots + \frac{y-y_{n-1}}{b_{1,n}}} + \cdots + \cfrac{x-x_{m-1}}{b_{m,0} + \frac{y-y_0}{b_{m,1}} + \frac{y-y_1}{b_{m,2}} + \cdots + \frac{y-y_{n-1}}{b_{m,n}}} \tag{15}$$

and $b_{i,j}(i = 0, 1, \ldots, m; j = 0, 1, \ldots, n)$ represent the bivariate partial inverse differences.

Theorem 3 ([1,2,38]). *If $b_{i,j}(i = 0, 1, \ldots, m; j = 0, 1, \ldots, n)$ exists, then*

$$R(x_i, y_j) = f_{i,j}, \forall (x_i, y_j) \in \prod_{m,n}, i = 0, 1, \ldots, m; j = 0, 1, \ldots, n. \tag{16}$$

3.1. Bivariate Thiele-Type Branched Rational Interpolation Continued Fractions with a Single Parameter

By introducing new parameters λ ($\lambda \neq 0$), an arbitrary point of the original points $(x_k, y_l, f_{k,l})(k = 0, 1, \ldots, m; l = 0, 1, \ldots, n)$ is treated as a virtual double point, and the multiplicity of the other points remains the same. We can construct the bivariate Thiele-like branched rational interpolation continued fractions with a single parameter λ using the following Algorithm 1:

Algorithm 1 Algorithm of the bivariate Thiele-like branched rational interpolation continued fractions with a single parameter

Step 1: Initialization.
$$f_{i,j}^{(0,0)} = f(x_i, y_j), i = 0, 1, \ldots, m; j = 0, 1, \ldots, n. \tag{17}$$

Step 2: For $j = 0, 1, \ldots, n; p = 1, 2, \ldots, m; i = p, p+1, \ldots, m$,
$$f_{i,j}^{(p,0)} = \frac{x_i - x_{p-1}}{f_{i,j}^{(p-1,0)} - f_{p-1,j}^{(p-1,0)}}. \tag{18}$$

Step 3: For $i = 0, 1, \ldots, k-1, k+1, \ldots, m; q = 1, 2, \ldots, n; j = q, q+1, \ldots, n$,
$$f_{i,j}^{(i,q)} = \frac{y_j - y_{q-1}}{f_{i,j}^{(i,q-1)} - f_{i,q-1}^{(i,q-1)}}. \tag{19}$$

Step 4: By introducing parameter λ into the formula $f_{k,j}^{(k,0)}$ ($j = 0, 1, \ldots, n$), then one can calculate them with Formulas (2)–(5), and mark the final results as
$$(a_{k,0}, a_{k,1}, \ldots, a_{k,l}, a_{k,l+1}^0, a_{k,l+1}, a_{k,l+2}, \ldots, a_{k,n})^T. \tag{20}$$

Step 5: Using the elements in Formulas (19) and (20), the Thiele-like interpolation continued fractions with a single parameter with respect to y can be constructed:
$$A_i(y) = \begin{cases} f_{i,0}^{(i,0)} + \dfrac{y-y_0}{f_{i,1}^{(i,1)}} + \dfrac{y-y_1}{f_{i,2}^{(i,2)}} + \cdots + \dfrac{y-y_{n-1}}{f_{i,n}^{(i,n)}}, & i = 0, 1, \ldots, k-1, k+1, \ldots, m, \\ a_{k,0} + \dfrac{y-y_0}{a_{k,1}} + \cdots + \dfrac{y-y_{l-1}}{a_{k,l}} + \dfrac{y-y_l}{a_{k,l+1}^0} + \dfrac{y-y_l}{a_{k,l+1}} + \dfrac{y-y_{l+1}}{a_{k,l+2}} + \cdots + \dfrac{y-y_{n-1}}{a_{k,n}}, & i = k, \end{cases} \tag{21}$$

Step 6: Let
$$R_{m,n}^0(x,y) = A_0(y) + \frac{x - x_0}{A_1(y)} + \frac{x - x_1}{A_2(y)} + \cdots + \frac{x - x_{m-1}}{A_m(y)}. \tag{22}$$

Then, $R_{m,n}^0(x,y)$ is a bivariate Thiele-like branched rational interpolation continued fractions with a single parameter.

Theorem 4. *Given the interpolation data* $(x_i, y_j, f_{i,j})$ $(i = 0, 1, \ldots, m; j = 0, 1, \ldots, n)$, *the bivariate Thiele-like branched rational interpolation continued fractions with a single parameter* $R_{m,n}^0(x,y)$ *satisfies*
$$R_{m,n}^0(x_i, y_j) = f_{i,j}, \forall (x_i, y_j) \in \prod_{m,n}, i = 0, 1, \ldots, m; j = 0, 1, \ldots, n. \tag{23}$$

Proof. For an arbitrary point $(x_i, y_j, f_{i,j})$, $i = 0, 1, \ldots, k-1, k+1, \ldots, m$, obviously
$$A_i(y_j) = f_{i,0}^{(i,0)} + \frac{y_j - y_0}{f_{i,1}^{(i,1)}} + \frac{y_j - y_1}{f_{i,2}^{(i,2)}} + \cdots + \frac{y_j - y_{t-1}}{f_{i,j}^{(i,j)}} = \cdots = f_{i,j}^{(i,0)}.$$

If $i = k$,
$$A_i(y) = a_{k,0} + \frac{y-y_0}{a_{k,1}} + \cdots + \frac{y-y_{l-1}}{a_{k,l}} + \frac{y-y_l}{a_{k,l+1}^0} + \frac{y-y_l}{a_{k,l+1}} + \frac{y-y_{l+1}}{a_{k,l+2}} + \cdots + \frac{y-y_{n-1}}{a_{k,n}},$$

Regardless of $0 \leq j < l, j = l, n \geq j > l$, from Theorem 1 in the Thiele-like rational interpolation continued fractions with a single parameter [33,34], we can derive

$$A_i(y_j) = a_{k,0} + \frac{y_j - y_0}{a_{k,1}} + \cdots + \frac{y_j - y_{l-1}}{a_{k,l}} + \frac{y_j - y_l}{a_{k,l+1}^0} + \frac{y_j - y_l}{a_{k,l+1}} + \frac{y_j - y_{l+1}}{a_{k,l+2}} + \cdots + \frac{y_j - y_{j-1}}{a_{k,j}}$$
$$= \cdots = f_{i,j}^{(i,0)},$$

From Theorem 3, we can derive, $\forall (x_s, y_t) \in \prod_{m,n}$,

$$R_{m,n}^0(x_i, y_j) = A_0(y_j) + \frac{x_i - x_0}{A_1(y_j)} + \frac{x_i - x_1}{A_2(y_j)} + \cdots + \frac{x_i - x_{i-1}}{A_i(y_j)}$$
$$= f_{0,j}^{(0,0)} + \frac{x_i - x_0}{f_{1,j}^{(1,0)}} + \frac{x_i - x_1}{f_{2,j}^{(2,0)}} + \cdots + \frac{x_i - x_{i-1}}{f_{i,j}^{(i,0)}}$$
$$= \cdots = f_{i,j}^{(0,0)} = f_{i,j}.$$

Then, we have proved the Theorem 4. □

3.2. Bivariate Thiele-Like Branched Rational Interpolation Continued Fractions with Multiple Parameters

Without loss of generality, we just develop the Thiele-like branched interpolation continued fractions with two parameters, which can be divided into three cases: one is taking a point as a virtual treble point, one is taking two virtual double points in the same column, and the other is taking two virtual double points in the different columns. The bivariate Thiele-like branched rational interpolation continued fractions with more than two parameters can be discussed similarly.

3.2.1. Bivariate Thiele-Like Branched Rational Interpolation Continued Fractions with Two Parameters Based on a Virtual Treble Point

By introducing new parameters α, β ($\alpha \neq 0, \beta \neq 0$), an arbitrary point of the original point $(x_k, y_l, f_{k,l})$ ($k = 0, 1, \ldots, m; l = 0, 1, \ldots, n$) is regarded as a treble virtual point, and the multiplicity of the other points remains the same. We can construct the bivariate Thiele-like rational interpolation continued fractions with two parameters α, β using Algorithm 2:

Algorithm 2 Algorithm of the bivariate Thiele-like rational interpolation continued fractions with two parameters

Step 1: Initialization:
$$f_{i,j}^{(0,0)} = f(x_i, y_j), i = 0, 1, \ldots, m; j = 0, 1, \ldots, n. \tag{24}$$

Step 2: If $j = 0, 1, \ldots, n; p = 1, 2, \ldots, m; i = p, p+1, \ldots, m$,
$$f_{i,j}^{(p,0)} = \frac{x_i - x_{p-1}}{f_{i,j}^{(p-1,0)} - f_{p-1,j}^{(p-1,0)}}. \tag{25}$$

Step 3: For $i = 0, 1, \ldots, k-1, k+1, \ldots, m; q = 1, 2, \ldots, n; j = q, q+1, \ldots, n$,
$$f_{i,j}^{(i,q)} = \frac{y_j - y_{q-1}}{f_{i,j}^{(i,q-1)} - f_{i,q-1}^{(i,q-1)}}. \tag{26}$$

By introducing parameters α, β into the formula $f_{k,j}^{(k,0)}$ ($j = 0, 1, \ldots, n$), then one can calculate the final result as

$$(a_{k,0}, a_{k,1}, \ldots, a_{k,l}, a_{k,l+1}^0, a_{k,l+1}^1, a_{k,l+1}, a_{k,l+2}, \ldots, a_{k,n})^T. \tag{27}$$

Step 4: By using the elements in Formulas (26) and (27), the Thiele-like interpolation continued fractions with a single parameter with respect to y can be constructed:

$$A_i(y) = \begin{cases} f_{i,0}^{(i,0)} + \frac{y-y_0}{f_{i,1}^{(i,1)}} + \frac{y-y_1}{f_{i,2}^{(i,2)}} + \cdots + \frac{y-y_{n-1}}{f_{i,n}^{(i,n)}}, & i = 0, 1, \ldots, k-1, k+1, \ldots, m, \\ a_{k,0} + \frac{y-y_0}{a_{k,1}} + \cdots + \frac{y-y_{l-1}}{a_{k,l}} + \frac{y-y_l}{a_{k,l+1}^0} + \frac{y-y_l}{a_{k,l+1}^1} + \frac{y-y_l}{a_{k,l+1}} + \frac{y-y_l}{a_{k,l+2}} + \cdots + \frac{y-y_{n-1}}{a_{k,n}}, & i = k. \end{cases} \tag{28}$$

Step 5: Let
$$R_{m,n}^1(x, y) = A_0(y) + \frac{x - x_0}{A_1(y)} + \frac{x - x_1}{A_2(y)} + \cdots + \frac{x - x_{m-1}}{A_m(y)}. \tag{29}$$

Then, $R_{m,n}^1(x, y)$ is a bivariate Thiele-like branched rational interpolation continued fractions with two parameters based on a treble virtual point.

Theorem 5. *Given the interpolation data* $(x_i, y_j, f_{i,j})$ $(i = 0, 1, \ldots, m; j = 0, 1, \ldots, n)$, *the bivariate Thiele-like branched rational interpolation continued fractions with two parameters based on a treble virtual point $R_{m,n}^1(x, y)$ satisfies*

$$R_{m,n}^1(x_i, y_j) = f_{i,j}, \forall (x_i, y_j) \in \prod_{m,n}, i = 0, 1, \ldots, m; j = 0, 1, \ldots, n. \tag{30}$$

We can prove Theorem 5 by using Theorem 3 and the method similar to the Theorem 1 in the Thiele-like rational interpolation continued fractions with a single parameter [33,34].

3.2.2. Bivariate Thiele-Like Branched Rational Interpolation Continued Fractions with Two Parameters Based on Two Virtual Double Points in the Same Column

Similar to the univariate Thiele-like interpolation continued fractions, we can get the bivariate Thiele-like branched rational interpolation continued fractions with two parameters. By introducing new parameters ϕ, δ ($\phi \neq 0, \delta \neq 0$), two arbitrary points of the original points $(x_k, y_l, f_{k,l}), (x_k, y_s, f_{k,s})$ ($s > l, k = 0, 1, \ldots, m; l, s = 0, 1, \ldots, n$) are treated as two virtual double points, and the multiplicity of the other points remains the same. One can construct the Thiele-like branched rational interpolation continued fractions with parameters ϕ, δ based on two virtual double points in the same column using Algorithm 3:

Algorithm 3 Algorithm of the Thiele-like branched rational interpolation continued fractions with two parameters in the same column

Step 1: Initialization:
$$f_{i,j}^{(0,0)} = f(x_i, y_j), i = 0, 1, \ldots, m; j = 0, 1, \ldots, n. \tag{31}$$

Step 2: If $j = 0, 1, \ldots, n; p = 1, 2, \ldots, m; i = p, p+1, \ldots, m$,
$$f_{i,j}^{(p,0)} = \frac{x_i - x_{p-1}}{f_{i,j}^{(p-1,0)} - f_{p-1,j}^{(p-1,0)}}. \tag{32}$$

Step 3: For $i = 0, 1, \ldots, k-1, k+1, \ldots, m; q = 1, 2, \ldots, n; j = q, q+1, \ldots, n$,
$$f_{i,j}^{(i,q)} = \frac{y_j - y_{q-1}}{f_{i,j}^{(i,q-1)} - f_{i,q-1}^{(i,q-1)}}. \tag{33}$$

Step 4: By introducing parameters ϕ, δ into $f_{k,j}^{(k,0)}$ ($j = 0, 1, \ldots, n$), then one can calculate them with inverse differences similar to Equation (2)–(5), and mark the final results as
$$\left(a_{k,0}, a_{k,1}, \ldots, a_{k,l}, a_{k,l+1}^0, a_{k,l+1}, a_{k,l+2}, \ldots, a_{k,s}, a_{k,s+1}^0, a_{k,s+1}, a_{k,s+2}, \ldots, a_{k,n}\right)^T. \tag{34}$$

Step 5: Using the elements in Formulas (52) and (53), the univariate Thiele-like interpolation continued fractions with two parameters with respect to y can be constructed:

$$A_i(y) = \begin{cases} f_{i,0}^{(i,0)} + \frac{y-y_0}{f_{i,1}^{(i,1)}} + \frac{y-y_1}{f_{i,2}^{(i,2)}} + \cdots + \frac{y-y_{n-1}}{f_{i,n}^{(i,n)}}, & i = 0, 1, \ldots, k-1, k+1, \ldots, m, \\ a_{k,0} + \frac{y-y_0}{a_{k,1}} + \cdots + \frac{y-y_{l-1}}{a_{k,l}} + \frac{y-y_l}{a_{k,l+1}^0} + \frac{y-y_l}{a_{k,l+1}} + \frac{y-y_{l+1}}{a_{k,l+2}} + \cdots + \\ \frac{y-y_{s-1}}{a_{k,s}} + \frac{y-y_s}{a_{k,s+1}^0} + \frac{y-y_s}{a_{k,s+1}} + \frac{y-y_{s+1}}{a_{k,s+2}} + \cdots + \frac{y-y_{n-1}}{a_{k,n}}, & i = k, \end{cases} \tag{35}$$

Step 6: Let
$$R_{m,n}^2(x, y) = A_0(y) + \frac{x - x_0}{A_1(y)} + \frac{x - x_1}{A_2(y)} + \cdots + \frac{x - x_{m-1}}{A_m(y)}. \tag{36}$$

Then, $R_{m,n}^2(x, y)$ is a bivariate Thiele-type branched interpolation continued fraction with two parameters based on two virtual double nodes in the same column.

Theorem 6. *Given the interpolation data $(x_i, y_j, f_{i,j})$ $(i = 0, 1, \ldots, m; j = 0, 1, \ldots, n)$, the bivariate Thiele-like branched rational interpolation continued fractions with two parameters based on two double virtual nodes in same column $R_{m,n}^2(x, y)$ satisfies*

$$R_{m,n}^2(x_i, y_j) = f_{i,j}, \forall (x_i, y_j) \in \prod_{m,n}, i = 0, 1, \ldots, m; j = 0, 1, \ldots, n. \tag{37}$$

We can prove Theorem 6 by using the method similar to the Theorem 1 in the Thiele-like rational interpolation continued fractions with a single parameter [33,34].

3.2.3. Bivariate Thiele-Like Branched Rational Interpolation Continued Fractions with Two Parameters Based on Two Virtual Double Points in Different Columns

By introducing new parameters φ, ω ($\varphi \neq 0, \omega \neq 0$), two arbitrary points of the original points $(x_k, y_l, f_{k,l}), (x_s, y_t, f_{s,t})$ ($s \neq k, t \neq l, s, k = 0, 1, \ldots, m; l, t = 0, 1, \ldots, n$) are treated as two virtual double points, and the multiplicity of the other points remains the same. One can construct the Thiele-like rational interpolation continued fractions with parameters φ, ω based on two virtual double points on the different columns using Algorithm 4:

Algorithm 4 Algorithm of the Thiele-like rational interpolation continued fractions with two parameters on the different columns

---

Step 1: Initialization.
$$f_{i,j}^{(0,0)} = f(x_i, y_j), i = 0, 1, \ldots, m; j = 0, 1, \ldots, n. \tag{38}$$

Step 2: If $j = 0, 1, \ldots, n; p = 1, 2, \ldots, m; i = p, p+1, \ldots, m$,
$$f_{i,j}^{(p,0)} = \frac{x_i - x_{p-1}}{f_{i,j}^{(p-1,0)} - f_{p-1,j}^{(p-1,0)}}. \tag{39}$$

Step 3: For $i = 0, 1, \ldots, k-1, k+1, \ldots, s-1, s+1, \ldots, m; q = 1, 2, \ldots, n; j = q, q+1, \ldots, n$,
$$f_{i,j}^{(i,q)} = \frac{y_j - y_{q-1}}{f_{i,j}^{(i,q-1)} - f_{i,q-1}^{(i,q-1)}}. \tag{40}$$

Step 4: By introducing parameter φ into $f_{k,j}^{(k,0)}$ ($j = 0, 1, \ldots, n$), we can calculate them by using Formulas (2)–(5) and mark the final results as
$$\left(a_{k,0}, a_{k,1}, \ldots, a_{k,l}, a_{k,l+1}^0, a_{k,l+1}, a_{k,l+2}, \ldots, a_{k,n}\right)^T. \tag{41}$$

Step 5: By introducing parameter ω into $f_{s,j}^{(s,0)}$ ($j = 0, 1, \ldots, n$), we can calculate them by using Formulas (2)–(5), and mark the final results as
$$\left(a_{s,0}, a_{s,1}, \ldots, a_{s,t}, a_{s,t+1}^0, a_{s,t+1}, a_{s,t+2}, \ldots, a_{s,n}\right)^T. \tag{42}$$

Step 6: By using the elements in Formulas (40)–(42), the Thiele-like interpolation continued fractions with a single parameter with respect to y was constructed:

$$A_i(y) = \begin{cases} f_{i,0}^{(i,0)} + \dfrac{y-y_0}{f_{i,1}^{(i,1)}} + \dfrac{y-y_1}{f_{i,2}^{(i,2)}} + \cdots + \dfrac{y-y_{n-1}}{f_{i,n}^{(i,n)}}, & i = 0,1,\ldots,k-1, k+1,\ldots,s-1, s+1,\ldots,m, \\ a_{k,0} + \dfrac{y-y_0}{a_{k,1}} + \cdots + \dfrac{y-y_{l-1}}{a_{k,l}} + \dfrac{y-y_l}{a_{k,l+1}^0} + \dfrac{y-y_l}{a_{k,l+1}} + \dfrac{y-y_{l+1}}{a_{k,l+2}} + \cdots + \dfrac{y-y_{n-1}}{a_{k,n}}, & i = k, \\ a_{s,0} + \dfrac{y-y_0}{a_{s,1}} + \cdots + \dfrac{y-y_{t-1}}{a_{s,t}} + \dfrac{y-y_t}{a_{s,t+1}^0} + \dfrac{y-y_t}{a_{s,t+1}} + \dfrac{y-y_{t+1}}{a_{s,t+2}} + \cdots + \dfrac{y-y_{n-1}}{a_{s,n}}, & i = s, \end{cases} \tag{43}$$

Step 7: Let
$$R_{m,n}^3(x, y) = A_0(y) + \frac{x - x_0}{A_1(y)} + \frac{x - x_1}{A_2(y)} + \cdots + \frac{x - x_{m-1}}{A_m(y)}. \tag{44}$$

Then, $R_{m,n}^3(x, y)$ is a bivariate Thiele-like branched rational interpolation continued fractions with two parameters based on two virtual double points in the different columns.

Theorem 7. Given the interpolation data $(x_i, y_j, f_{i,j})$ $(i = 0, 1, \ldots, m; j = 0, 1, \ldots, n)$, the bivariate Thiele-like branched rational interpolation continued fractions with two parameters based on two virtual double points in different columns $R^3_{m,n}(x, y)$ satisfies

$$R^3_{m,n}(x_i, y_j) = f_{i,j}, \forall (x_i, y_j) \in \prod_{m,n}, \ i = 0, 1, \ldots, m; j = 0, 1, \ldots, n. \tag{45}$$

We can prove Theorem 7 by using Theorem 3 and the method similar to the Theorem 1 in the Thiele-like rational interpolation continued fractions with a single parameter [33,34].

3.3. Dual Bivariate Thiele-Like Branched Rational Interpolation Continued Fractions with a Single Parameter

It is easy to see that the new interpolation methods were computed with respect to the partial inverse difference of x firstly, and then with respect to the partial inverse difference of y from Algorithms 1–4. In fact, we can also do that with respect to the partial inverse difference of y and then with respect to the partial inverse difference of x. By introducing a new parameter θ ($\theta \neq 0$), an arbitrary point of the original points $(x_k, y_l, f_{k,l})$ ($k = 0, 1, \ldots, m; l = 0, 1, \ldots, n$) is treated as a virtual double point, and taking Algorithm 3 as an example, the multiplicity of the other points remains the same. One can construct the Thiele-like branched rational interpolation continued fractions with parameters θ ($\theta \neq 0$) based on this virtual double point using Algorithm 5:

Algorithm 5 Algorithm of the dual bivariate Thiele-like branched rational interpolation continued fractions with a single parameter

---

Step 1: Initialization:

$$f^{(0,0)}_{i,j} = f(x_i, y_j), i = 0, 1, \ldots, m; \ j = 0, 1, \ldots, n. \tag{46}$$

Step 2: If $i = 0, 1, \ldots, m; q = 1, 2, \ldots, n; j = q, q+1, \ldots, n$,

$$f^{(0,q)}_{i,j} = \frac{y_j - y_{q-1}}{f^{(0,q-1)}_{i,j} - f^{(0,q-1)}_{i,q-1}}. \tag{47}$$

Step 3: For $j = 0, 1, \ldots, l-1, l+1, \ldots, n; p = 1, 2, \ldots, m; i = p, p+1, \ldots, m$,

$$f^{(p,j)}_{i,j} = \frac{x_i - x_{p-1}}{f^{(p-1,j)}_{i,j} - f^{(p-1,j)}_{p-1,j}}. \tag{48}$$

Step 4: By introducing a parameter θ into $f^{(0,l)}_{i,l}$ ($i = 0, 1, \ldots, m$), we can calculate them using Formulas (2)–(5) and mark the final results as

$$a_{0,l}, a_{1,l}, \ldots, a_{k,l}, a^0_{k+1,l}, a_{k+1,l}, a_{k+2,l}, \ldots, a_{m,l}. \tag{49}$$

Step 5: By using the elements in Formulas (48) and (49), the Thiele-type interpolation continued fractions with a single parameter regarded to x can be constructed:

$$A_j(x) = \begin{cases} f^{(0,j)}_{0,j} + \dfrac{x - x_0}{f^{(1,j)}_{1,j}} + \dfrac{x - x_1}{f^{(2,j)}_{2,j}} + \cdots + \dfrac{x - x_{m-1}}{f^{(m,j)}_{m,j}}, & j = 0, 1, \ldots, l-1, l+1, \ldots, n, \\ a_{0,l} + \dfrac{x - x_0}{a_{1,l}} + \cdots + \dfrac{x - x_{k-1}}{a_{k,l}} + \dfrac{x - x_k}{a^0_{k+1,l}} + \dfrac{x - x_k}{a_{k+1,l}} + \dfrac{x - x_{k+1}}{a_{k+2,l}} + \cdots + \dfrac{x - x_{m-1}}{a_{m,l}}, & j = l, \end{cases} \tag{50}$$

Step 6: Let

$$R^4_{m,n}(x, y) = A_0(x) + \frac{y - y_0}{A_1(x)} + \frac{y - y_1}{A_2(x)} + \cdots + \frac{y - y_{n-1}}{A_n(x)}. \tag{51}$$

Then, $R^4_{m,n}(x, y)$ is a dual bivariate Thiele-like branched rational interpolation continued fractions with a single parameter.

Theorem 8. *Given the interpolation data* $(x_i, y_j, f_{i,j})$ $(i = 0, 1, \ldots, m; j = 0, 1, \ldots, n)$, *the dual bivariate Thiele-like branched rational interpolation continued fractions with a single parameter* $R^4_{m,n}(x, y)$ *satisfies*

$$R^4_{m,n}(x_i, y_j) = R^0_{m,n}(x_i, y_j) = f_{i,j}, \forall (x_i, y_j) \in \prod_{m,n}, i = 0, 1, \ldots, m; j = 0, 1, \ldots, n. \tag{52}$$

Proof. For an arbitrary point $(x_i, y_j, f_{i,j})$, it is easy to prove

$$R^0_{m,n}(x_i, y_j) = f_{i,j}.$$

If $j = 0, 1, \ldots, l-1, l+1, \ldots, n$, obviously, we have

$$A_j(x_i) = f^{(0,j)}_{0,j} + \frac{x_i - x_0}{f^{(1,j)}_{1,j}} + \frac{x_i - x_1}{f^{(2,j)}_{2,j}} + \cdots + \frac{x_i - x_{i-1}}{f^{(i,j)}_{i,j}} = \cdots = f^{(0,j)}_{i,j}.$$

If $j = l$,

$$A_j(x) = a_{0,l} + \frac{x - x_0}{a_{1,l}} + \cdots + \frac{x - x_{k-1}}{a_{k,l}} + \frac{x - x_k}{a^0_{k+1,l}} + \frac{x - x_k}{a_{k+1,l}} + \frac{x - x_{k+1}}{a_{k+2,l}} + \cdots + \frac{x - x_{m-1}}{a_{m,l}}.$$

Regardless of $0 \leq i < k, i = k, n \geq i > k$, from the Theorem 1 in the Thiele-like rational interpolation continued fractions with a single parameter [33,34], we have

$$A_j(x_i) = a_{0,l} + \frac{x_i - x_0}{a_{1,l}} + \cdots + \frac{x_i - x_{k-1}}{a_{k,l}} + \frac{x_i - x_k}{a^0_{k+1,l}} + \frac{x_i - x_k}{a_{k+1,l}} + \frac{x_i - x_{k+1}}{a_{k+2,l}} + \cdots + \frac{x_i - x_{i-1}}{a_{i,l}} = \cdots = f^{(0,j)}_{i,j}.$$

So, we have

$$\begin{aligned} R^4_{m,n}(x_i, y_j) &= A_0(x_i) + \frac{y_j - y_0}{A_1(x_i)} + \frac{y_j - y_1}{A_2(x_i)} + \cdots + \frac{y_j - y_{j-1}}{A_j(x_i)} \\ &= f^{(0,0)}_{i,0} + \frac{y_j - y_0}{f^{(0,1)}_{i,1}} + \frac{y_j - y_1}{f^{(0,2)}_{i,2}} + \cdots + \frac{y_j - y_{j-1}}{f^{(0,j)}_{i,j}} \\ &= \cdots = f^{(0,0)}_{i,j} = f_{i,j}. \end{aligned}$$

Then, we can obtain the result. □

We call $R^4_{m,n}(x, y)$ as the dual interpolation of $R^0_{m,n}(x, y)$. In addition, we can also study many dual bivariate Thiele-like branched rational interpolation continued fractions with two or more parameters, similar to the discussion in Section 3.3.

According to the process of the various new rational interpolation, it can be easily seen that every new Thiele-like rational interpolation continued fractions with parameters has many special interpolations which enables proper parameters to be selected. The advantages of the proposed methods are easy to compute, have adjustable parameters, deal with unattainable points, and so on. Different interpolation functions can be derived according to their own practical needs. Meanwhile, the novel interpolation functions can be adjusted at an arbitrary point in the interpolant region under unaltered interpolant data by selecting appropriate parameters, so the interpolation curves or surfaces were modified. However, it is still difficult to select appropriate parameters and then construct a proper interpolation function for meeting the practical geometric design requirement and the need for image interpolant processing and other related problems. We will study the geometric design and image interpolation based on the new Thiele-like interpolation continued fractions with parameters in the future.

3.4. Bivariate Thiele-Like Branched Rational Interpolation Continued Fractions with Unattainable Points

Definition 2. *Given the point set $D = \{(x_i, y_j, f_{i,j}) | i = 0, 1, \ldots, m, j = 0, 1, \ldots, n\}$, where (x_i, y_j) is diverse, $R(x, y) = \frac{N(x,y)}{D(x,y)}$ is the bivariate Thiele-like branched rational interpolation continued fractions in Formula (15) if the point $(x_k, y_l, f_{k,l})$ satisfies*

$$N(x_k, y_l) - f_{k,l} \cdot D(x_k, y_l) = 0, R(x_k, y_l) = \frac{N(x_k, y_l)}{D(x_k, y_l)} \neq f_{k,l}. \tag{53}$$

Then, $(x_k, y_l, f_{k,l})$ is regarded as an unattainable point of $R(x, y)$.

Theorem 9 ([31]). *Suppose the bivariate Thiele-like branched rational interpolation continued fractions which is diverse for the given point set $D = \{(x_i, y_j, f_{i,j}) | i = 0, 1, \ldots, m, j = 0, 1, \ldots, n\}$, shown in Formula (15), satisfies*

$$a_{ij} \neq \infty (i = 0, 1, \ldots, m-1, j = 0, 1, \ldots, n-1), a_{mj} \neq 0, a_{in} \neq 0, i = 1, 2, \ldots, m; j = 1, 2, \ldots, n, \tag{54}$$

then the necessary and sufficient condition of $(x_k, y_l, f_{k,l})$ is an unattainable point where $A_k^{(j)}(y_j) = 0$, and $A_k^{(s)}(y) = \frac{N_k^{(s)}(y)}{D_k^{(s)}(y)}$ is irreducible, and

$$A_k^{(n-1)}(y) = a_{i,n}, A_k^{(s)}(y) = a_{k,s+1} + (y - y_{s+1})/A_k^{(s+1)}(y), s = n-2, n-3, \ldots, j. \tag{55}$$

Theorem 10. *The bivariate Thiele-like rational interpolation branched continued fraction with a single parameter $R_{m,n}^0(x, y)$ by Algorithm 3 satisfies*

$$R_{m,n}^0(x_i, y_j) = f_{i,j}, \forall (x_i, y_j) \in \prod_{m,n}, i = 0, 1, \ldots, m; j = 0, 1, \ldots, n. \tag{56}$$

This makes the unattainable point change into an accessible point.

The proof method is similar to Theorem 2.

3.5. Bivariate Thiele-Like Branched Rational Interpolation Continued Fractions for Nonexistent Partial Inverse Difference

Similar to the univariate Thiele-like rational interpolation continued fractions with a nonexistent inverse difference, when we construct bivariate Thiele-like branched rational interpolation continued fractions for the diverse interpolation data on the given rectangular net, $\{(x_0, y_0), (x_1, y_1), \ldots, (x_n, y_n)\}$, it may meet the interpolation problem that the partial inverse difference does not exist. In this case, we can adjust the multiple numbers of interpolation points. We also can solve this problem by using the bivariate Thiele-type branched rational interpolation continued fractions with one or two parameters or the dual bivariate Thiele-like branched rational interpolation continued fractions with one or more parameters.

4. Numerical Examples

In this section, we provide some examples to illustrate how this method is implemented and its flexibility. The first example is given to demonstrate that the Thiele-like interpolation continued fractions with parameters are stable for the Runge function. The second example shows the interpolation with unattainable points in classical Thiele-type continued fractions rational interpolation. The third example is the multivariate Thiele-like rational interpolation continued fractions with parameters.

To enrich the application of the proposed algorithm, we present an image zoom example based on parameterized bivariate Thiele-like rational interpolation continued fractions in the fourth example.

Example 1. *For the function* $f(x) = \frac{1}{1+25x^2}$, *the higher-degree polynomial interpolation is unstable.*

We can derive the classic Newton polynomial interpolation with the given data at points $-1, -0.8, -0.6, -0.4, -0.2, 0$:

$$P = 0.03846 + 0.1018(x+1) + 0.26025(x+1)(x+0.8) + 0.7916667(x+1)(x+0.8)(x+0.6)$$
$$+ 2.686979(x+1)(x+0.8)(x+0.6)(x+0.4) - 6.36354(x+1)(x+0.8)(x+0.6)(x+0.4)(x+0.2)$$

We can get the Thiele-type continued fractions interpolation:

$$R = \frac{1923}{5000} + \frac{x+1}{9.823182711} + \frac{x+0.8}{-0.06018033} + \frac{x+0.6}{-37.753208} + \frac{x+0.4}{0.021847219} + \frac{x+0.2}{-5883.58062}.$$

We can use Thiele-like interpolation continued fractions with parameters to calculate it. As the function has symmetry, we just discuss the condition within $[-1, 0]$ of the interpolation interval. If we set $[-0.8, 0.05882]$ as a virtual double point, we can get

$$R = \frac{1923}{5000} + \frac{x+1}{\frac{509}{5000}} + \frac{x+0.8}{c} + \frac{x+0.8}{5(c + \frac{1566193}{26025000})} + \frac{x+0.6}{\frac{-(26025000c + 1566193)(62787500c + 4111193)}{50000(32680893750c + 1793617763)}}$$

$$+ \frac{x+0.4}{\frac{-(24(32680893750c + 1793617763)(27478062500c + 1461475921))}{2545(26025000c + 1566193)(16895932679c + 956527052)}}$$

$$+ \frac{x+0.2}{\frac{(16895932679c + 956527052)(844806165185c + 47828425248)}{8(32680893750c + 1793617763)(1689901538395c + 95660995792)}}$$

As shown in Table 2, the Thiele-like interpolation continued fractions can perform better than the Newton polynomial interpolation with six points. However, these two methods cannot interpolate all of the interpolation data. The Thiele-like rational interpolation continued fractions with a parameter invariably satisfies the interpolation condition with different values of parameter c, and gives a better effect, and we also can get some new Thiele-like rational interpolation continued fractions formulas by selecting parameters. This is similar to the block-based Thiele-like blending rational interpolation [30]. We also can construct many other Thiele-type rational interpolation continued fractions using other virtual points or many parameters.

Table 2. Comparison table of different interpolation.

x_i	$f(x_i)=\frac{1}{1+25x_i^2}$	Newton Interpolation $p_6(x)$	Thiele Continued Fractions Interpolation	Thiele Continued Fractions Interpolation with $c = 1$	Thiele Continued Fractions Interpolation with $c = -10$
−1.00	0.03846	0.03846	0.03846000000	0.03846000000	0.03846000000
−0.96	0.04160	0.03298	0.04159488287	0.04159595868	0.04159595865
−0.90	0.04706	0.03770	0.04705547745	0.04705768836	0.04705768826
−0.86	0.05131	0.04534	0.05130475006	0.05130739205	0.05130739185
−0.80	0.05882	0.05882	0.05882000000	0.05882000000	0.05882000000
−0.76	0.06477	0.06763	0.06476367286	0.06476628712	0.06476628741
−0.70	0.07547	0.07971	0.07546947663	0.07547137553	0.07547137561
−0.66	0.08410	0.08737	0.08410290267	0.08410410847	0.08410410852
−0.60	0.10000	0.10000	0.09999999823	0.10000000000	0.10000000000
−0.56	0.11312	0.11071	0.11312302677	0.11312226113	0.11312226113
−0.50	0.13793	0.13338	0.13793269272	0.13793118870	0.13793118867
−0.46	0.15898	0.15487	0.15898409183	0.15898264914	0.15898264914
−0.40	0.2000	0.20000	0.19999998642	0.20000000000	0.20000000000
−0.36	0.23585	0.24042	0.23584678820	0.23584889034	0.23584889036
−0.30	0.30769	0.31883	0.30768550203	0.30769184706	0.30769184712
−0.26	0.37175	0.38372	0.37173857114	0.37174665711	0.37174665718
−0.20	0.50000	0.50000	0.49999990866	0.50000000000	0.50000000000
−0.16	0.60976	0.73720	0.60977966699	0.60975745619	0.60975745603
−0.10	0.80000	0.73720	0.80009647464	0.80000519086	0.80000519027
−0.06	0.91743	0.84193	0.91756721089	0.91743821176	0.91743821096
0.00	1.0000	1.00000	0.99999979851	1.00000000000	1.00000000000

Example 2. *Given the following interpolation data in Table 3, the corresponding inverse difference table of Thiele-type continued fractions interpolation is shown in Tables 4 and 5.*

Table 3. Interpolation data.

i	0	1	2
x_i	2	1	0
f_i	1	0	0

Table 4. Table of inverse differences.

Nodes	0 Order Inverse Differences	1 Order Inverse Differences	2 Order Inverse Differences
2	1		
1	0	1	
0	0	2	-1

Table 5. Table of inverse differences with a parameter.

Nodes	0 Order Inverse Differences	1 Order Inverse Differences	2 Order Inverse Differences	3 Order Inverse Differences
2	1			
2	1	λ		
1	0	1	$\frac{\frac{1}{\lambda}-1}{\lambda-1}$	
0	0	2	$\frac{\frac{2}{\lambda}}{\lambda-2}$	$3-\frac{2}{\lambda}-\lambda$

So, we can get the Thiele-type continued fractions rational interpolation:

$$r(x) = 1 + \frac{x-2}{1} + \frac{x-1}{-1} = 0.$$

As $r(x_0) = r(2) = 0 \neq 1$, we can see that $(2,1)$ is a unique unattainable point. Following the algorithm in this paper, by adding multiple numbers of node $(2,1)$, the osculating interpolation which has a first-order derivative at point $(2,1)$, introducing parameter $\lambda (\lambda \neq 0)$, and constructing the inverse difference table shown in Table 5 above, we gain the corresponding Thiele-type osculatory rational interpolation:

$$R_2(x) = 1 + \frac{x-2}{\lambda} + \frac{x-2}{\frac{1}{\lambda-1}} + \frac{x-1}{3-\frac{2}{\lambda}-\lambda}.$$

It is easy to verify that

$$R_2(x_i) = f_i \quad (i = 0,1,2).$$

So, by choosing the different value of parameter λ, the function $R_2(x)$ invariably satisfies the given interpolation data. Meantime, this method can well solve this kind of special interpolation problem, and it is easy to construct and calculate. In addition, the function $R_2(x)$ can be converted into many other rational interpolation functions. For example, we can choose the following functions: If we choose $\lambda = -3$, we can get

$$R_2(x) = 1 + \frac{x-2}{-3} + \frac{x-2}{-\frac{1}{4}} + \frac{x-1}{\frac{20}{3}}.$$

If we choose $\lambda = 80$, we can get

$$R_2(x) = 1 - \frac{x-2}{-80} + \frac{x-2}{\frac{40x}{3081} - \frac{1}{39}}. \tag{57}$$

As can be seen from Figure 1, both functions satisfy the interpolation condition. We can choose other values of the parameters $\lambda (\lambda \neq 0)$, and the function $R_3(x)$ can change into other functions.

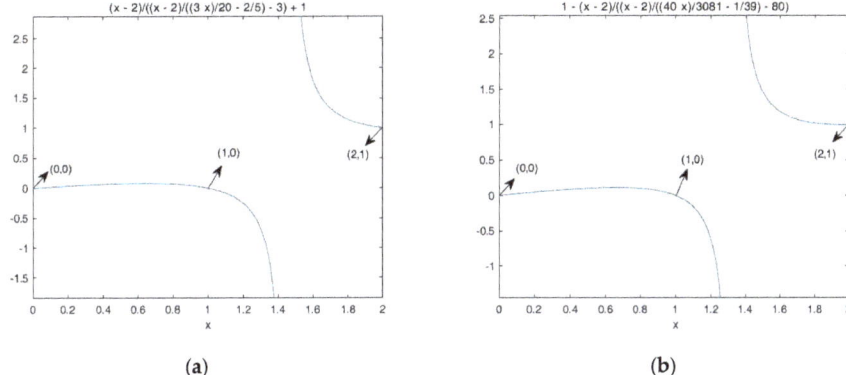

Figure 1. (a) Graph of $R_2(x)$ with $\lambda = -3$, (b) graph of $R_2(x)$ with $\lambda = 80$.

Example 3. *The interpolation data are given in Table 6.*

Table 6. Interpolation data.

(x_i, y_j)	$x_0=0$	$x_1=0.5$	$x_2=1$
$y_0 = 0$	2	2.3	2.5
$y_1 = 0.5$	1.8	2	2.1
$y_1 = 1$	1.5	1.55	1.5

We can get the bivariate Thiele-type blending rational interpolation using the method presented in [1]:

$$R(x,y) = 2 - \frac{y}{\frac{5}{2} + \frac{y-0.5}{-1}} + \frac{x}{\frac{5}{3} + \frac{y}{3 + \frac{y-0.5}{-\frac{25}{24}}}} + \frac{x-0.5}{\frac{3}{2} + \frac{y}{-\frac{5}{9} + \frac{y-0.5}{-\frac{9}{2}}}}.$$

Following Algorithm 3 in this paper, we add the multiplicity of point $(0,0,2)$ and construct the osculating interpolation which has its first-order derivative at point $(0,0,2)$ by introducing parameter λ:

$$R^{(0)}(x,y) = 2 + \frac{y}{\lambda + \frac{y}{\frac{-1}{2\lambda+5} + \frac{y-0.5}{\frac{-(2\lambda+5)(\lambda+2)}{2(\lambda+3)}}}} + \frac{x}{\frac{5}{3} + \frac{y}{3 + \frac{y-0.5}{-\frac{25}{24}}}} + \frac{x-0.5}{\frac{3}{2} + \frac{y}{-\frac{5}{9} + \frac{y-0.5}{-\frac{9}{2}}}}.$$

It is easy to verify that the function $R_2(x)$ invariably satisfies the given interpolation data with the different value of parameter λ, i.e.,

$$R(x_i, y_j) = R^{(0)}(x_i, y_j) = f_{ij}(i, j = 0, 1, 2).$$

We can modify the bivariate blending rational interpolation by selecting parameters, but $R(x,y)$ cannot, so our method gives a new choice for the application, and it gives a new method for studying the rational interpolation theory.

Example 4. *Image interpolation is an important method in pixel level image processing, the interpolated data are often regarded as a certain interpolation kernel and a linear combination of the input image pixels in traditional methods. Due to the influence of light, natural background and image texture characteristics, generally speaking, the adjacent pixels of an image are not a simple linear relationship. In order to obtain more effective and better visual results, many nonlinear methods have been proposed for the image interpolation in the literature. To enrich the application of the proposed parameterized bivariate Thiele-like rational interpolation* continued fractions algorithm, we take an image zoom as an example. We choose $\lambda = 1$, the performances of the proposed parameterized Thiele-like continued fraction rational interpolation method can be deduced from image interpolation process. In our experiment, we take the image "Lenna" as the test image as shown in Figure 2. The original image is resized by a factor 2 (see Figure 3) with four image interpolation methods. The experiment results demonstrate that the zoomed images do not have obvious jagged edges with the proposed parameterized Thiele-like continued fraction rational interpolation method, so the proposed algorithm can be used for image interpolation processing effectively. It is obvious that our new method is implemented without producing the so-called mosaics or blocky effect, and the results maintain clearness of the image, including edges, and the details are maintained well, hence, it offers more detailed information. From Figure 4, we can see, when the image is enlarged by a larger factor, the new proposed algorithm still has better visual performance.

Figure 2. The original Lenna Image.

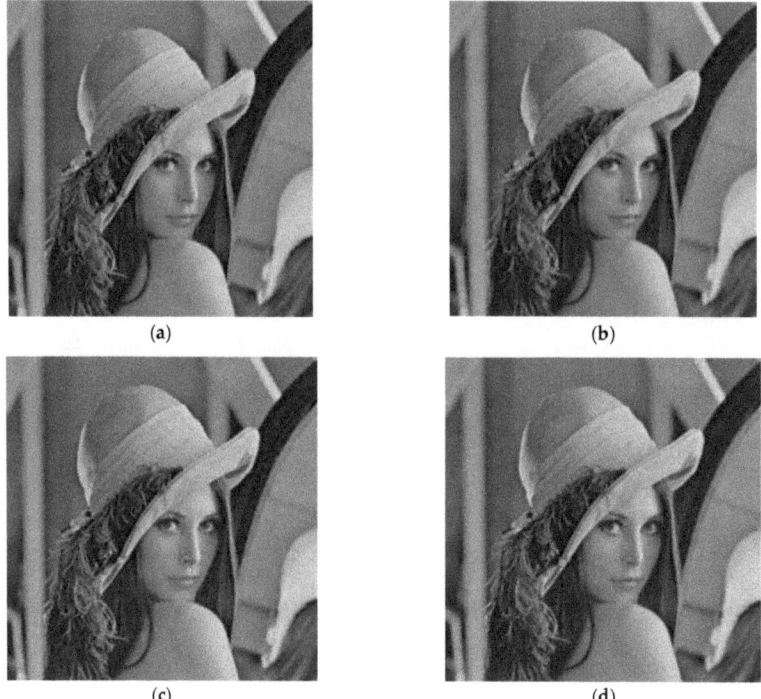

Figure 3. The zoomed images by factor 2 based on four different interpolant methods. (**a**) The zoomed images by factor 2 based on the nearest-neighbor interpolation method, (**b**) the zoomed images by factor 2 based on the bilinear interpolation, (**c**) the zoomed images by factor 2 based on the bicubic interpolation, (**d**) the zoomed images by factor 2 based on the proposed method.

(a)   (b)

Figure 4. Comparison of eye effects of the nearest neighbor interpolation and the proposed rational interpolation. (a) The zoomed images by factor 4 based on the nearest-neighbor interpolation method, (b) the zoomed images by factor 4 based on the proposed method.

5. Conclusions and Future Work

In this paper, by strategically selecting the multiplicity of the interpolation nodes, we have developed many types of univariate and bivariate Thiele-like rational interpolation continued fractions with parameters. We also discussed the interpolant algorithms, interpolant theorems, and dual interpolation. The new kinds of Thiele-like rational interpolation continued fractions are easy to use and extend the theoretical research and application of the rational interpolation functions. The value of the interpolation function can be changed at any point in the interpolant region under unaltered interpolant points by selecting appropriate parameters; therefore, it can be used to design a curve or surface. Based on the geometric design needs, it can alter the shape of curves or surfaces to satisfy actual needs. However, it is still a complicated problem. The selection of proper parameters and construction of an appropriate interpolation method for actual geometric design requirements are a very practical and interesting problem, and we will study it in the future. Further research on the following aspects will be summarized in our next study:

- How to select appropriate parameters and suitably alter the shape of the curves or surfaces according to actual requirements.
- The geometric properties of curves/surfaces based on the Thiele-like rational interpolation continued fractions function with parameters.
- How to design geometric modeling using Thiele-like rational interpolation continued fractions functions with parameters.
- The proposed Thiele-like rational interpolation continued fractions with parameters algorithm can be implemented in other pixel level image processing, such as image inpainting, removal of salt and pepper noise, image rotation, image super-resolution reconstruction, image metamorphosis and image upscaling.
- How to generalize the proposed algorithms to lacunary rational interpolants, rational interpolants over triangular grids.

To conclude, by using of the Sampson generalized inverse, it is not complicated to generalize the Thiele-like rational interpolation continued fractions algorithms with parameters to vector-valued cases or matrix-valued cases [1,2].

Author Contributions: Writing—original draft, methodology, L.Z.; supervision, writing—review and editing, L.S. and X.W.; validation, Y.C. and C.Z.; formal analysis, C.T. All authors have read and agreed to the published version of the manuscript.

Funding: The authors would like to express their thanks to the referees for their valuable suggestions. This work was supported by the grant from the National Natural Science Foundation of China, Nos. 61672204, 61806068; in part by the Key Scientific Research Foundation of Education Department of Anhui Province, Nos. KJ2018A0555, KJ2018A0556, KJ2019A0833; the Natural Science Foundation of Anhui Provincial, Nos. 1908085MF184, 1508085QF116; in part by the Key Technologies R&D Program of Anhui Province, No. 1804a09020058; in part by the Major Science and Technology Project of Anhui Province, No. 17030901026; and in part by the Key Constructive Discipline Project of Hefei University, No. 2016xk05.

Conflicts of Interest: The authors declare no conflict of interest. The funders had no role in the design of the study; in the collection, analyses, or interpretation of data; in the writing of the manuscript, or in the decision to publish the results.

References

1. Tan, J. *Theory of Continued Fractions and Its Applications*; Science Publishers: Beijing, China, 2007.
2. Wang, R.; Zhu, G. *Approximation of Rational Interpolation and Its Application*; Science Publishers: Beijing, China, 2004.
3. Kuchmins'ka, K.; Vonza, S. On Newton-Thiele-like Interpolating Formula. *Commun. Anal. Theory Contin. Fractions* **2000**, *8*, 74–79.
4. Pahirya, M.; Svyda, T. Problem of Interpolation of Functions by Two-Dimensional Continued Fractions. *Ukrai. Math. J.* **2006**, *58*, 954–966. [CrossRef]
5. Li, S.; Dong, Y. Viscovatov-Like Algorithm of Thiele-Newton's Blending Expansion for a Bivariate Function. *Mathematics* **2019**, *7*, 696. [CrossRef]
6. Cuyt, A.; Celis, O. Multivariate Data Fitting With Error Control. *Bit Numer. Math.* **2019**, *59*, 35–55. [CrossRef]
7. Li, G.; Zhang, X.; Yang, H. Numerical Analysis, Circuit Simulation, And Control Synchronization of Fractional-Order Unified Chaotic System. *Mathematics* **2019**, *7*, 1077. [CrossRef]
8. Xu, Q.; Liu, Z. Scattered Data Interpolation and Approximation with Truncated Exponential Radial Basis Function. *Mathematics* **2019**, *7*, 1101. [CrossRef]
9. Massopust, P. Non-Stationary Fractal Interpolation. *Mathematics* **2019**, *7*, 666. [CrossRef]
10. Akal, C.; Lukashov, A. Newton-Padé Approximations for Multivariate Functions. *Appl. Math. Comput.* **2018**, *334*, 367–374. [CrossRef]
11. Cuyt, A.; Verdonk, B. General Order Newton-Padé Approximants for Multivariate Functions. *Numer. Math.* **1984**, *43*, 293–307. [CrossRef]
12. Ravi, M.S. Geometric Methods in Rational Interpolation Theory. *Linear Algebra Appl.* **1997**, *258*, 159–168. [CrossRef]
13. Bertrand, F.; Calvi, J. The Newton Product of Polynomial Projectors Part 1: Construction and Algebraic Properties. *Int. J. Math.* **2019**, *30*, 1950030. [CrossRef]
14. Akal, C.; Lukashov, A. Scale of Mean Value Multivariate Padé Interpolations. *Filomat* **2017**, *31*, 1123–1128. [CrossRef]
15. Li, H.B.; Song, M.Y.; Zhong, E.J.; Gu, X.M. Numerical Gradient Schemes for Heat Equations Based on the Collocation Polynomial and Hermite Interpolation. *Mathematics* **2019**, *7*, 93. [CrossRef]
16. Yu, K.; Wang, C.; Yang, S.; Lu, Z.; Zhao, D. An Effective Directional Residual Interpolation Algorithm for Color Image Demosaicking. *Appl. Sci.* **2018**, *8*, 680. [CrossRef]
17. Zhou, X.; Wang, C.; Zhang, Z.; Fu, Q. Interpolation Filter Design Based on All-Phase DST And Its Application to Image Demosaicking. *Information* **2018**, *9*, 206. [CrossRef]
18. Min, H.; Jia, W.; Zhao, Y. LATE: A Level Set Method Based on Local Approximation of Taylor Expansion for Segmenting Intensity Inhomogeneous Images. *IEEE Trans. Image Process.* **2018**, *27*, 5016–5031. [CrossRef]
19. He, L.; Tan, J.; Xing, Y.; Hu, M.; Xie, C. Super-Resolution Reconstruction Based on Continued Fractions Interpolation Kernel in The Polar Coordinates. *J. Electron. Imaging* **2018**, *27*, 043035. [CrossRef]
20. He, L.; Tan, J.; Su, Z.; Luo, X.; Xie, C. Super-resolution by polar Newton-Thiele's rational kernel in centralized sparsity paradigm. *Signal Process. Image Commun.* **2015**, *31*, 86–99. [CrossRef]
21. Yao, X.; Zhang, Y.; Bao, F.; Liu, F.; Zhang, M. The Blending Interpolation Algorithm Based on Image Features. *Multimed. Tools Appl.* **2018**, *77*, 1971–1995. [CrossRef]

22. Zhang, Y.; Fan, Q.; Bao, F.; Liu, Y.; Zhang, C. Single-Image Super-Resolution Based on Rational Fractal Interpolation. *IEEE Trans. Image Process.* **2018**, *27*, 3782–3797.
23. Ahn, H.E.; Jeong, J.; Kim, J.W.; Kwon, S.; Yoo, J. A Fast 4K Video Frame Interpolation Using a Multi-Scale Optical Flow Reconstruction Network. *Symmetry* **2019**, *11*, 1251. [CrossRef]
24. Wei, X.; Wu, Y.; Dong, F.; Zhang, J.; Sun, S. Developing an Image Manipulation Detection Algorithm Based on Edge Detection and Faster R-CNN. *Symmetry* **2019**, *11*, 1223. [CrossRef]
25. Zhang, Y.; Bao, F.; Zhang, M. A Rational Interpolation Surface Model and Visualization Constraint. *Sci. Sin. Math.* **2014**, *44*, 729–740. [CrossRef]
26. Zhang, Y.; Bao, X.; Zhang, M.; Duan, Q. A Weighted Bivariate Blending Rational Interpolation Function and Visualization Control. *J Comput. Anal. Appl.* **2012**, *14*, 1303–1320.
27. Zou, L.; Tang, S. New Approach to Bivariate Blending Rational Interpolants. *Chin. Q. J. Math.* **2011**, *26*, 280–284.
28. Zou, L.; Tang, S. General Structure of Block-Based Interpolational Function. *Commun. Math. Res.* **2012**, *28*, 193–208.
29. Zou, L.; Tang, S. A New Approach to General Interpolation Formulae for Bivariate Interpolation. *Abstr. Appl. Anal.* **2014**, *2014*, 421635. [CrossRef]
30. Zhao, Q.; Tan, Q. Block-based Thiele-like Blending Rational Interpolation. *J. Comput. Appl. Math.* **2006**, *195*, 312–325. [CrossRef]
31. Zhu, X.; Zhu, G. A Study of the Existence of Vector Valued Rational Interpolation. *J. Inf. Comput. Sci.* **2005**, *2*, 631–640.
32. Zhu, X. Research and Application of Rational Function Interpolation. Ph.D. Thesis, University of Science and Technology of China, Hefei, China, 2002.
33. Zou, L.; Song, L.; Wang, X.; Huang, Q.; Chen, Y.; Tang, C.; Zhang, C. Univariate Thiele Type Continued Fractions Rational Interpolation with Parameters. In *International Conference on Intelligent Computing*; Springer: Cham, Switzerland, 2019; pp. 399–410.
34. Huang, D.; Huang, Z.; Hussain, A. *Intelligent Computing Methodologies*; Springer Science and Business Media LLC: Berlin/Heidelberg, Germany, 2019.
35. Li, C.W.; Zhu, X.L.; Pan, Y.L. A Study of The Unattainable Point for Rational Interpolation. *Coll. Math.* **2010**, *26*, 50–55.
36. Zhao, Q.; Tan, Q. Successive Newton-Thiele's Rational Interpolation. *J. Inf. Comput. Sci.* **2005**, *2*, 295–301.
37. Zhao, Q.; Tan, Q. Block-based Newton-like Blending Rational Interpolation. *J. Comput. Math.* **2006**, *24*, 515–526.
38. Siemazko, W. Thiele-type Branched Continued Fractions for Two Variable Functions. *J. Comput. Appl. Math.* **1983**, *9*, 137–153. [CrossRef]

© 2020 by the authors. Licensee MDPI, Basel, Switzerland. This article is an open access article distributed under the terms and conditions of the Creative Commons Attribution (CC BY) license (http://creativecommons.org/licenses/by/4.0/).

Article

Mixed Generalized Multiscale Finite Element Method for Darcy-Forchheimer Model

Denis Spiridonov [1], Jian Huang [2,3,4], Maria Vasilyeva [5,6,*], Yunqing Huang [2,3,4] and Eric T. Chung [7]

[1] Multiscale Model Reduction Laboratory, North-Eastern Federal University, 677980 Yakutsk, Republic of Sakha (Yakutia), Russia; d.stalnov@mail.ru
[2] School of Mathematics and Computational Science, Xiangtan University, Xiangtan 411105, China; huangjian213@xtu.edu.cn (J.H.); huangyq@xtu.edu.cn (Y.H.)
[3] Hunan Key Laboratory for Computation and Simulation in Science and Engineering, Xiangtan 411105, China
[4] Key Laboratory of Intelligent Computing Information Processing of Ministry of Education, Xiangtan 411105, China
[5] Institute for Scientific Computation, Texas A&M University, College Station, TX 77843-3368, USA
[6] Department of Computational Technologies, North-Eastern Federal University, 677980 Yakutsk, Republic of Sakha (Yakutia), Russia
[7] Department of Mathematics, The Chinese University of Hong Kong (CUHK), Hong Kong, China; tschung@math.cuhk.edu.hk
* Correspondence: vasilyevadotmdotv@gmail.com

Received: 7 October 2019; Accepted: 5 December 2019; Published: 10 December 2019

Abstract: In this paper, the solution of the Darcy-Forchheimer model in high contrast heterogeneous media is studied. This problem is solved by a mixed finite element method (MFEM) on a fine grid (the reference solution), where the pressure is approximated by piecewise constant elements; meanwhile, the velocity is discretized by the lowest order Raviart-Thomas elements. The solution on a coarse grid is performed by using the mixed generalized multiscale finite element method (mixed GMsFEM). The nonlinear equation can be solved by the well known Picard iteration. Several numerical experiments are presented in a two-dimensional heterogeneous domain to show the good applicability of the proposed multiscale method.

Keywords: Darcy-Forchheimer model; flow in porous media; nonlinear equation; heterogeneous media; finite element method; multiscale method; mixed generalized multiscale finite element method; multiscale basis functions; two-dimensional domain

1. Introduction

The Darcy-Forchheimer equation is commonly used for describing the high velocity flow near oil and gas wellbores and fractures, which is a correction formula of the well known Darcy's law by supplementing a nonlinear velocity quantity as follows:

$$\mu k^{-1} \boldsymbol{u} + \beta \rho \, |\boldsymbol{u}| \, \boldsymbol{u} + \nabla p = 0, \tag{1}$$

where μ, k, ρ and β represent the viscosity, the permeability, the density, and the dynamic viscosity coefficient of the fluid, respectively. β is also mentioned as the Forchheimer coefficient, whose values stand for the nonlinear intensity. In contrast, Darcy's law, which is valid for the extremely small velocity case, is usually used to show the linear relationship between the velocity vector $\boldsymbol{u}$ and the pressure gradient ∇p. The Darcy-Forchheimer model can be obtained by coupling Equation (1) with the following conservation law equation:

$$\nabla \cdot u = f. \tag{2}$$

In recent years, the Darcy-Forchheimer model has been studied by many researchers within numerical discretized methods. Girault et al. in [1] proved the existence and uniqueness of the solution of the Darcy-Forchheimer model. Then, they considered mixed finite element methods by piecewise constant and nonconforming Crouzeix-Raviart elements to approximate the velocity and the pressure, respectively. Park in [2] gave a mixed finite element method (MFEM) for generalized Darcy-Forchheimer flow. Pan et al. in [3] presented an MFEM to approximate the velocity based on the Raviart–Thomas element or the Brezzi–Douglas–Marini element and piecewise constant for the pressure of the Darcy-Forchheimer model. Rui et al. in [4] published a block-centered finite difference method (BCFDM) for the Darcy-Forchheimer model. The authors in [5] established a BCFDM for the Darcy-Forchheimer model with variable parameter $\beta(x)$. Rui and Liu in [6] introduced a two level BCFDM for the Darcy-Forchheimer model. Huang, Chen, and Rui in [7] designed a nonlinear multigrid method for the two-dimensional Darcy-Forchheimer model with a Peaceman–Rachford-type iteration as a smoother and proposed a better choice of the parameter used in the splitting. We point out that the constructed multigrid method with an almost linear computational cost is convergent independent of the critical parameters.

To solve the problem in heterogeneous media, a very fine mesh should be used for solving a heterogeneity scale problem, which results to a large number of degrees of freedom. For dimension reduction and fast solvers of such problems, some model reduction techniques are needed. In [8], the authors introduced the mixed multiscale finite element methods and presented the main convergence results for the solution of second order elliptic equations with heterogeneous coefficients, which oscillate rapidly. Mixed multiscale finite element methods are widely used for solving the reservoir simulation problems [9–11]. In [12–14], we developed a mixed generalized multiscale finite element method (GMsFEM), where we enriched a multiscale space by new degrees of freedom, which was obtained by solving local spectral problems on the snapshot space.

In this paper, we consider a solution of the Darcy-Forchheimer model in heterogeneous media. We construct an efficient algorithm of the mixed generalized multiscale finite element method to make an approximation of the problem on the coarse grid. The construction is based on solving the local problems for calculating multiscale basis functions of the velocity. Meanwhile, we use piecewise constant elements to approximate the pressure. In the mixed formulation, we firstly define a snapshot space, which provides a solution space in each local area. Then we solve the local spectral problem in the snapshot space to find out multiscale basis functions. We use the Picard iteration to address the non-linearity when solving the problems on the multiscale spaces [15,16]. Note that when constructing multiscale basis functions, we do not take into account the nonlinear part of the Darcy-Forchheimer equation.

The remainder of this article is organized as follows: The model problem and its weak formulation in mixed form and the discrete weak formulation are demonstrated in Section 2. The fine grid approximation and coarse gird approximation are presented in Section 3 and Section 4, respectively. Some numerical experiments using our mixed generalized multiscale finite element method are carried out in Section 5 to verify the efficiency of the presented method. Finally, conclusions and further ideas are presented in Section 6.

2. Mathematical Model

We consider the steady state Darcy-Forchheimer model to describe a single phase fluid flow in a heterogeneous porous medium.

$$\mu k^{-1}(x)u + \beta(x)\rho|u|u + \nabla p = 0, \quad x \in \Omega,$$
$$\nabla \cdot u = f, \quad x \in \Omega, \quad (3)$$

where $\Omega \in \mathbb{R}^2$ is a bounded domain and $\partial \Omega$ is Lipschitz continuous; $|v| = (v,v)^{\frac{1}{2}}$, $(\cdot,\cdot)$ denotes the L^2 inner product; k and β are the heterogeneous permeability and the heterogeneous non-Darcy coefficient, respectively.

Complemented with Dirichlet boundary condition,

$$p = f_D, \quad x \in \partial \Omega. \quad (4)$$

Without loss of generality, we suppose that $f_D = 0$, namely the homogeneous Dirichlet boundary condition. Then, we get the following problem:

$$\mu k^{-1}(x)u + \beta(x)\rho|u|u + \nabla p = 0, \quad x \in \Omega,$$
$$\nabla \cdot u = f, \quad x \in \Omega, \quad (5)$$
$$p = 0, \quad x \in \partial \Omega.$$

Remark 1. *The boundary condition (4) can be replaced by the Neumann boundary condition,*

$$u \cdot n = f_N, \quad x \in \partial \Omega,. \quad (6)$$

f and f_N should satisfy the compatibility condition by the Gauss theorem,

$$\int_\Omega f(x) \, dx = \int_{\partial \Omega} f_N(s) \, ds. \quad (7)$$

3. Fine Grid Approximation

In this section, mixed finite element methods have been borrowed to handle Problem (5) in the fine grid. Here, we use the standard Sobolev spaces notation to define the function spaces and their norms as follows:

$$V = \left\{ v \in L^3(\Omega)^2; \nabla \cdot v \in L^2(\Omega) \right\}, \quad \|u\|_V = \|v\|_{0,3,\Omega} + \|\nabla \cdot v\|_{0,2,\Omega}, \quad \forall v \in V.$$

$$Q = L^2(\Omega), \quad \|q\|_Q = \|q\|_{0,2,\Omega}, \quad \forall q \in Q.$$

Then, we obtain the following variational formulation: find $(u,p) \in V \times Q$ such that:

$$\int_\Omega \mu k^{-1}(x) u \, v \, dx + \int_\Omega \beta(x)\rho|u|u \, v \, dx - \int_\Omega p \nabla \cdot v \, dx = 0, \quad \forall v \in V,$$
$$-\int_\Omega q \nabla \cdot u \, dx = -\int_\Omega f q \, dx, \quad \forall q \in Q. \quad (8)$$

The variational formulation (8) and the problem (5) are equivalent by the following Green's formula:

$$\int_\Omega \nabla p \, v \, dx = -\int_\Omega p \nabla \cdot v \, dx + \int_{\partial \Omega} p v \cdot n \, ds, \quad \forall v \in V. \quad (9)$$

Let Ω be a polygon in two dimensions, which can be entirely covered by a shape regular decomposition $\mathcal{T}_h$, in the sense of Ciarlet [17], into triangles, with h being the maximum diameter of the elements of the triangles. Therefore, $\mathcal{T}_h$ is a family of conforming triangulations of $\overline{\Omega}$,

$$\overline{\Omega} = \bigcup_{T \in \mathcal{T}_h} T.$$

Here, we discretize the velocity u in the lowest order Raviart–Thomas space:
Given a simplex $T \in \mathbb{R}^2$, the local Raviart–Thomas space of order $k \geq 0$ is defined by:

$$RT_k(T) = \mathcal{P}_k(T)^2 + x\,\mathcal{P}_k(T),$$

where $\mathcal{P}_k$ denotes the set of all polynomials in two variables of degree $\leq k$. Then, we can get the lowest order global Raviart–Thomas space, which is the conforming finite element space of the velocity u:

$$V_h = RT_0(\Omega) = \{v \in V : v|_T \in RT_0(T) \quad \forall\, T \in \mathcal{T}_h\}. \tag{10}$$

The pressure p is approximated in the following piecewise constant space:

$$Q_h = \left\{ q \in L^2(\Omega) : q|_T \in \mathcal{P}_0 \quad \forall\, T \in \mathcal{T}_h \right\}. \tag{11}$$

Then, we can obtain the discrete weak formulation of (8): find a pair $(u_h, p_h) \in V_h \times Q_h$:

$$\begin{aligned}
\int_\Omega \mu k^{-1}(x) u_h\, v_h\, dx + \int_\Omega \beta(x)\rho |u_h| u_h\, v_h\, dx - \int_\Omega p_h \nabla \cdot v_h\, dx &= 0, \quad \forall v_h \in V_h, \\
-\int_\Omega q_h \nabla \cdot u_h\, dx &= -\int_\Omega f\, q_h\, dx, \quad \forall q_h \in Q_h.
\end{aligned} \tag{12}$$

In [3], the authors demonstrated the existence and uniqueness of the continuous problem and the discrete problem, respectively. Moreover, if $\mathcal{T}_h$ is quasi-uniform and $(u, p) \in W^{s,3}(\Omega)^2 \times W^{s,\frac{3}{2}}(\Omega)$, then the following error estimates can be proven; see ([3], Theorem 4.4) for details:

$$\|u - u_h\|_{0,2}^2 + \|u - u_h\|_{0,3}^3 \leq C h^{2s}, \quad 1 \leq s \leq k+1, \tag{13}$$

$$\|p - p_h\|_{0,2} \leq C h^s, \quad 1 \leq s \leq k+1. \tag{14}$$

where $W^{s,p}(\Omega)$ is the standard Sobolev space of index (s, p), where s is a nonnegative integer and $p \geq 1$.

$$W^{s,p}(\Omega) = \{v \in L^p(\Omega) : D^\alpha v \in L^p(\Omega) \text{ for all } |\alpha| \leq s\},$$

where $D^\alpha v$ is the weak derivative of v; see the details in [18].

Let:

$$u_h = \sum_{i=1}^{m_1} u_i \xi_i, \quad p_h = \sum_{i=1}^{m_2} p_i \theta_i,$$

where $u = [u_1, u_2, \ldots, u_{m_1}]^T$, $p = [p_1, p_2, \ldots, p_{m_2}]^T$ are the coefficients of the finite element approximations with m_1, m_2 dimensions, in terms of a basis ξ of the velocity and a basis θ of the pressure, respectively.

We apply the Picard iteration for solving the resulting discrete nonlinear system.
Find $u_h^{n+1} \in V_h$, $p_h^{n+1} \in Q_h$ with an arbitrary initial guess $u_h^0 \in V_h$, such that:

$$\begin{aligned}
\int_\Omega \mu k^{-1}(x) u_h^{n+1} v_h\, dx + \int_\Omega \beta(x) \rho |u_h^n| u_h^{n+1} v_h\, dx - \int_\Omega p_h^{n+1} \nabla \cdot v_h\, dx &= 0, \quad \forall v_h \in V_h, \\
-\int_\Omega q_h \nabla \cdot u_h^{n+1}\, dx &= -\int_\Omega f\, q_h\, dx, \quad \forall q_h \in Q_h.
\end{aligned} \tag{15}$$

We rewrite the iteration (15) into the following matrix form.

$$\begin{pmatrix} A(u_h^n) & B^T \\ B & 0 \end{pmatrix} \begin{pmatrix} u_h^{n+1} \\ p_h^{n+1} \end{pmatrix} = \begin{pmatrix} 0 \\ F \end{pmatrix}, \tag{16}$$

where $A(u_h)$ is the matrix associated with the term:

$$\int_\Omega \mu k^{-1}(x) u_h v_h \, dx + \int_\Omega \beta(x) \rho |u_h| u_h v_h \, dx,$$

where B is the matrix corresponding to $-\int_\Omega q_h \nabla \cdot u_h \, dx$, and F is the vector associated with the linear functional $-\int_\Omega f q_h \, dx$.

In the practical implementation, we use the following termination criterion to control the iteration,

$$\max(r_u, r_p) < tol,$$

where:

$$r_u = \left\| \mu k^{-1}(x) u_h^{n+1} + \beta(x) \rho |u_h^{n+1}| u_h^{n+1} + \nabla p_h^{n+1} \right\|_0,$$

$$r_p = \begin{cases} \left\| f - \nabla \cdot u_h^{n+1} \right\|_0 / \|f\|_0, & \text{when } \|f\|_0 \neq 0, \\ \left\| f - \nabla \cdot u_h^{n+1} \right\|_0, & \text{when } \|f\|_0 = 0. \end{cases}$$

4. Coarse Grid Approximation

In this section, we describe the construction of the approximation on a coarse grid using the mixed generalized multiscale finite element method (mixed GMsFEM). We construct a square uniform coarse grid $\mathcal{T}_H$ for the computational domain Ω with a coarse grid size H; $\mathcal{E}_H = \cup_{i=1}^{N_E} E_i$ is the set of all facets of a coarse mesh, and N_E is the number of facets of a coarse mesh. To construct the multiscale basis function, we build a uniform triangular fine grid, which is obtained by refinement of a coarse grid. In the mixed GMsFEM, we compute the multiscale basis functions for the velocity in the local domains ω_i that correspond to the coarse edge $E_i \in \mathcal{E}_H$ (Figure 1).

$$\omega_i = \cup_j \{ K_j \in \mathcal{T}_H | E_i \subset \partial K_j \},$$

where K_j is the coarse grid cell, which is equal to a square element of $\mathcal{T}_H$.

We build a multiscale space for the velocity $u_{ms} \in V_{ms}$:

$$V_{ms} = \text{span}\{\psi_1, ..., \psi_{M_{\omega_i} \cdot N_E}\}, \tag{17}$$

where ψ_i are the multiscale basis functions, which are calculated in the local domain ω_i and M_{ω_i} is the number of basis functions in each local domain. For pressure, we use the space Q_{ms} of piecewise constant functions on the coarse cells.

We start with constructing a snapshot space in the local domain ω_i and then make an approximation on the coarse grid using the solution of the local spectral problem on the snapshot space. The snapshot space is obtained by solving the next local problem in ω_i: find $(\phi_j, \eta) \in V_h^{\omega_i} \times Q_h^{\omega_i}$ such that:

$$\begin{aligned} \int_{\omega_i} k^{-1} \phi_j^i v \, dx - \int_{\omega_i} \eta \nabla \cdot v \, dx &= 0, \quad v \in V_h^{\omega_i}, \\ \int_{\omega_i} r \nabla \cdot \phi_j^i dx &= \int_{\omega_i} c_j r \, dx, \quad r \in Q_h^{\omega_i}, \end{aligned} \tag{18}$$

with boundary condition:
$$\boldsymbol{\phi}_j^i \cdot \boldsymbol{n} = 0, \quad x \in \partial \omega_i, \tag{19}$$

where n is the unit exterior normal vector to the boundary $\partial \omega_i$.

On the fine facet e_j, we apply an additional boundary condition:
$$\boldsymbol{\phi}_j^i \cdot \boldsymbol{n} = \delta_{ij}, \quad x \in e_j, \tag{20}$$

where $j = 1, \ldots, J^{\omega_i}$ is the number of the fine facets e_j on which we define the boundary condition (20), $c_j = \frac{|e_j|}{S_{\omega_i}}$ is chosen by the compatibility condition $\int_{\partial \omega_i} \boldsymbol{\phi}_j^i \cdot \boldsymbol{n} = \int \nabla \cdot \boldsymbol{\phi}_j^i$, $|e_j|$ presents the length of the fine grid facet e_j, and S_{ω_i} denotes the volume of the local domain ω_i. Here, J^{ω_i} is the number of fine grid edges e_j on E_i, $E_i = \cup_{j=1}^{J^{\omega_i}} e_j$, and δ_{ij} is a piecewise constant function defined on E_i, which takes the value of one on e_j and zero on each other edge of the fine grid.

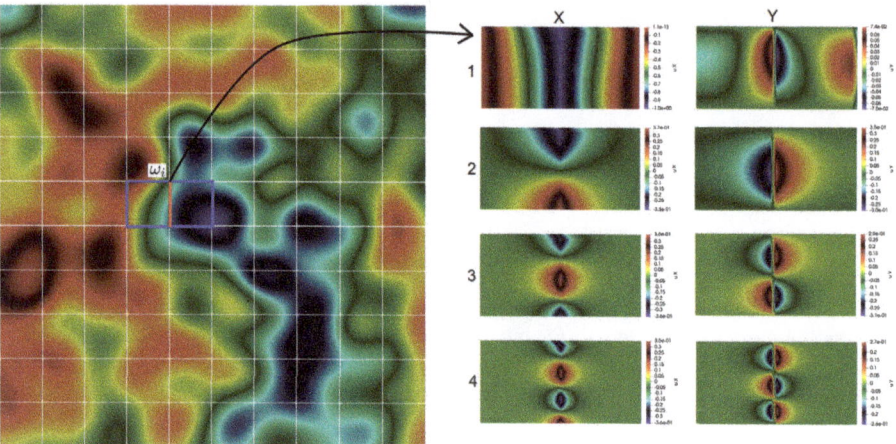

Figure 1. Illustration of the heterogeneous property, coarse grid, and multiscale basis functions in the local domain.

Therefore, we can define the snapshot space in each local domain ω_i:
$$V_{snap}^i = \text{span}\{\boldsymbol{\phi}_1^i, \ldots, \boldsymbol{\phi}_{J_i}^i\}.$$

To compute a multiscale basis function, we solve a local spectral problem in each local domain ω_i. The spectral problem allows us to find the most important characteristics of the problem. Then, we use multiscale basis functions to define the multiscale space to get an approximation on the coarse grid. In each ω_i, we solve the spectral problem on V_{snap}^i:
$$\tilde{A}^{\omega_i} \tilde{\boldsymbol{\psi}}_l^{\omega_i} = \lambda_l \tilde{S}^{\omega_i} \tilde{\boldsymbol{\psi}}_l^{\omega_i}, \quad \tilde{A}^{\omega_i} = R^{\omega_i} A^{\omega_i} (R^{\omega_i})^T, \quad \tilde{S}^{\omega_i} = R^{\omega_i} S^{\omega_i} (R^{\omega_i})^T, \tag{21}$$

where:
$$(R^{\omega_i})^T = [\boldsymbol{\phi}_1^i, \ldots, \boldsymbol{\phi}_{J^{\omega_i}}^i],$$

and:
$$A^{\omega_i} = [a_{mn}^{\omega_i}], \quad a_{mn}^{\omega_i} = a_{\omega_i}(\boldsymbol{\phi}_m, \boldsymbol{\phi}_n) = \int_{E_i} k^{-1} (\boldsymbol{\phi}_m \cdot \boldsymbol{n})(\boldsymbol{\phi}_n \cdot \boldsymbol{n}) \, ds, \tag{22}$$

$$S^{\omega_i} = [s_{mn}^{\omega_i}], \quad s_{mn}^{\omega_i} = s_{\omega_i}(\boldsymbol{\phi}_m, \boldsymbol{\phi}_n) = \int_{\omega_i} k^{-1} \boldsymbol{\phi}_m \boldsymbol{\phi}_n \, dx + \int_{\omega_i} \nabla \cdot \boldsymbol{\phi}_m \nabla \cdot \boldsymbol{\phi}_n \, dx. \tag{23}$$

For construction of the multiscale space, we select the first M_{ω_i} smallest eigenvalues and take the corresponding eigenvectors $\boldsymbol{\psi}_l^{\omega_i} = (R^{\omega_i})^T \tilde{\boldsymbol{\psi}}_l^{\omega_i}$ as basis functions ($l = 1, 2, \ldots, M_{\omega_i}$). We note that the presented spectral decomposition in the snapshot space is motivated by theoretical analysis (see Theorem 4.3 in [12]). Moreover, the oversampling techniques can be used for the construction of the multiscale basis functions to enhance the accuracy of mixed GMsFEM. The main idea of oversampling techniques is to introduce a small dimensional snapshot space using the POD (proper orthogonal decomposition) approach, where snapshot vectors are constructed in larger regions that contain the interfaces of two adjacent coarse blocks [12].

Remark 2. *For the construction of the multiscale basis functions, we can use other types of spectral problems (see [12,19]). For example, we can solve the following spectral problem:*

$$\tilde{S}^{\omega_i} \tilde{\psi}_l^{\omega_i} = \mu_l \tilde{\psi}_l^{\omega_i}, \quad \tilde{S}^{\omega_i} = R^{\omega_i} S^{\omega_i} (R^{\omega_i})^T, \tag{24}$$

with:

$$S^{\omega_i} = [s_{mn}^{\omega_i}], \quad s_{mn}^{\omega_i} = s_{\omega_i}(\boldsymbol{\phi}_m, \boldsymbol{\phi}_n) = \int_{\omega_i} k^{-1} \boldsymbol{\phi}_m \boldsymbol{\phi}_n \, dx, \tag{25}$$

on the snapshot space and taking eigenvectors corresponding to the largest eigenvalues as a multiscale basis functions ($\lambda_l = 1/\mu_l$).

Additionally, we calculate the first basis function by the solution of the following local problem in the domain ω_i: find $(\boldsymbol{\chi}^{\omega_i}, \eta) \in V_h^{\omega_i} \times Q_h^{\omega_i}$ such that:

$$\int_{\omega_i} k^{-1} \boldsymbol{\chi}^{\omega_i} \boldsymbol{v} \, dx - \int_{\omega_i} \eta \, \nabla \cdot \boldsymbol{v} \, dx = 0, \quad \boldsymbol{v} \in V_h^{\omega_i},$$

$$\int_{\omega_i} r \, \nabla \cdot \boldsymbol{\chi}^{\omega_i} \, dx = \int_{\omega_i} c \, r \, dx, \quad r \in Q_h^{\omega_i}, \tag{26}$$

with the following boundary conditions:

$$\boldsymbol{\chi}^{\omega_i} \cdot \boldsymbol{n} = 0, \quad x \in \partial \omega_i, \quad \boldsymbol{\chi}^{\omega_i} \cdot \boldsymbol{n} = 1, \quad x \in E_i, \tag{27}$$

where $c = \frac{|E_i|}{S_{\omega_i}}$, $|E_i|$ is the length of coarse facet E_i.

Finally, we obtain the following multiscale space for velocity:

$$V_{ms} = \text{span}\{\boldsymbol{\chi}^{\omega_i}, \boldsymbol{\psi}_l^{\omega_i}, 1 \leq l \leq M_{\omega_i}, 1 \leq i \leq N_E\}. \tag{28}$$

The illustration of the local multiscale basis functions with an additional basis are presented in Figure 1.

For the construction of the coarse grid system, we define a projection matrix:

$$R = \begin{bmatrix} R_u & 0 \\ 0 & R_p \end{bmatrix}, \quad R_u = [R_{u,1}, \ldots, R_{u,N_E}]^T, \tag{29}$$

where $(R_{u,i})^T = [\boldsymbol{\chi}^{\omega_i}, \boldsymbol{\psi}_1^{\omega_i}, \ldots, \boldsymbol{\psi}_{M_{\omega_i}}^{\omega_i}]$ and R_p is the projection matrix for pressure, where we set one for each fine grid cell in the current coarse grid cell. Here, N_E is the number of facets of the coarse grid, and M_{ω_i} is the number of multiscale basis functions in local domain ω_i.

We use the constructed multiscale space and write the approximation in the coarse grid in matrix form:

$$\begin{pmatrix} A_c^k & B_c^T \\ B_c & 0 \end{pmatrix} \begin{pmatrix} u_c^{k+1} \\ p_c^{k+1} \end{pmatrix} = \begin{pmatrix} 0 \\ F_c \end{pmatrix}, \tag{30}$$

where:

$$A_c^k = R_u A^k R_u^T, \quad B_c = R_u B R_p^T, \quad F_c = R_p F, \tag{31}$$

and finally, we reconstruct the solution on the fine grid $u_{ms}^{k+1} = R_u^T u_c^{k+1}$.

5. Numerical Results

In this section, we present several numerical results of the Darcy-Forchheimer model in the domain $\Omega = [0,1]^2$. We used a structured 160×160 fine grid with 77,120 edges and 51,200 cells (triangles) and a 10×10 coarse grid with 220 edges and 100 cells (see Figure 2).

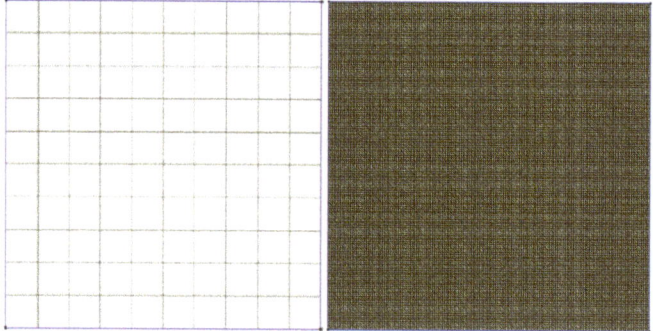

Figure 2. Coarse grid (**left**) and fine grid (**right**).

In the numerical study, we set $\mu = 1$, $\rho = 1$, and $f = 1$. For the Picard iteration, we set $tol = 10^{-8}$. The permeability tensor k is heterogeneous and presented in Figure 3, where we considered two test cases. We set the Darcy-Forchheimer coefficient $\beta = C \cdot k^{-1}$ from [20,21], where the parameter C controls the influence of the nonlinear part of the equation. We studied the proposed multiscale solver for $C = 10.24$, $C = 34.93$, $C = 1584.14$, and $C = 71,554.17$. With such values of C, we could investigate the behavior of a method with the various influences of the nonlinear part. By increasing the parameter C, we obtained a Darcy-Forchheimer equation with the dominant nonlinear part.

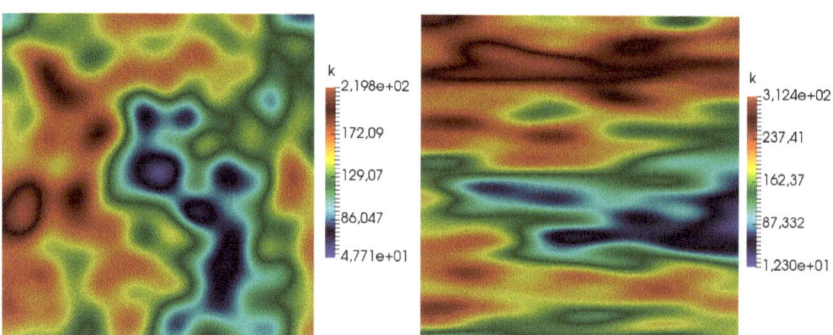

Figure 3. Heterogeneous coefficient for Test 1 (**left**) and Test 2 (**right**).

At first, we considered a test case with $\beta = 0$. Fine scale and multiscale solutions using eight multiscale basis functions are presented in Figures 4 and 5 for coefficient k from Test 1 and Test 2.

In Table 1, the relative errors in the L_2 norm for different numbers of multiscale basis functions are presented for $\beta = 0$. Here, DOF_c and DOF_f denote the size of the multiscale and fine grid solutions ($DOF_f = 128{,}320$), and M is the number of multiscale basis functions. By #iter, we denote the number of Picard iterations, and e_u and e_p are the relative errors in L_2 norm for velocity and pressure, respectively. To calculate the error of the pressure, the average values over the coarse grid cells are used.

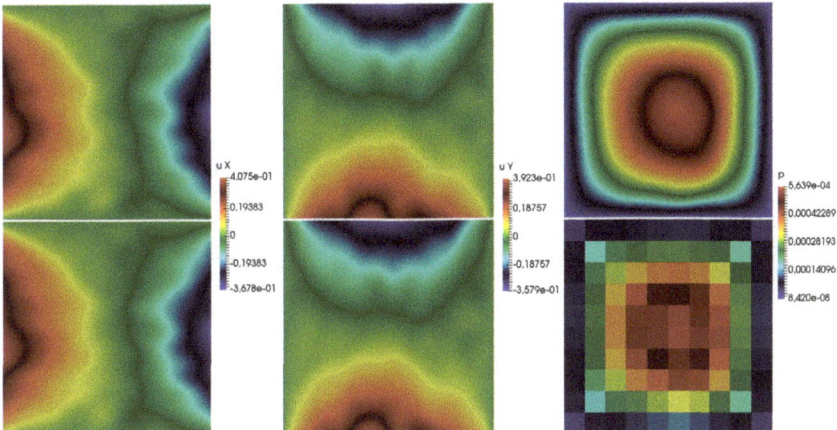

Figure 4. Fine grid solution (**top**) and multiscale solution using eight multiscale basis functions (**bottom**). Coefficient k from Test 1 with $\beta = 0$.

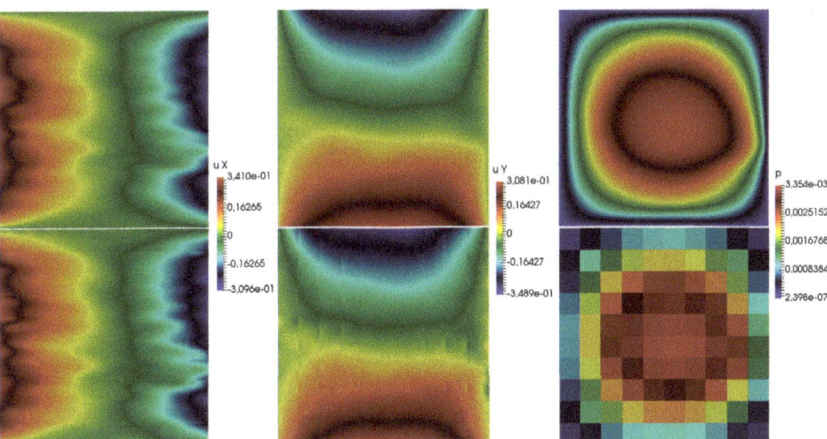

Figure 5. Fine grid solution (**top**) and multiscale solution using eight multiscale basis functions (**bottom**). Coefficient k from Test 2 with $C = 34.93$.

The numerical results for $C = 34.93$ with coefficient k from Test 1 and Test 2 using eight multiscale basis functions are presented in Figures 6 and 7. In Tables 2–5, we show the relative error in the L_2 norm between the multiscale solution and the fine grid solution for different numbers of multiscale basis functions. The errors are presented for different values of C to see the influence of the nonlinear part on the accuracy of mixed GMsFEM. According to the obtained results, we observed that the method worked well with the presented problem. When we increased the number of multiscale bases, we observed that the error decreased. For large values of C the error was greater than for smaller values. This was due to the fact that for large C values, the influence of the nonlinear part of the equation increased, and our multiscale bases did not take into account the nonlinear part. From the tables, we can observe that the number of Picard iterations of mixed GMsFEM was significantly smaller than the number of iterations of the fine grid solution for large C. For Test 1 and Test 2, we obtained similar results for the presented multiscale solver.

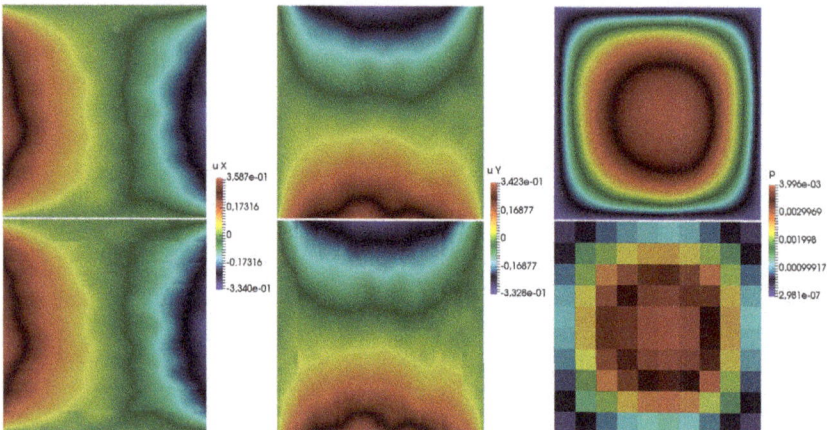

Figure 6. Fine grid solution (**top**) and multiscale solution using eight multiscale basis functions (**bottom**). Coefficient k from Test 1 with $C = 34.93$.

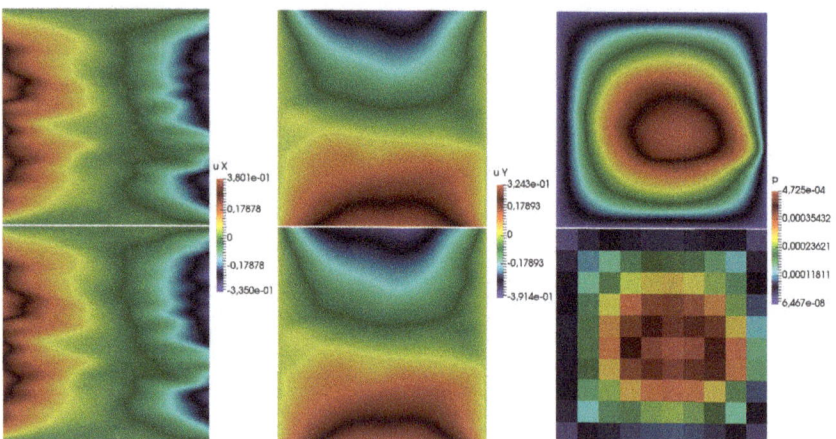

Figure 7. Fine grid solution (**top**) and multiscale solution using eight multiscale basis functions (**bottom**). Coefficient k from Test 2 with $\beta = 0$.

Table 1. Relative error in the L_2 norm for different numbers of multiscale basis functions. Coefficient k from Test 1 (left) and Test 2 (right) with $\beta = 0$.

M	DOF_c	e_u, %	e_p, %	M	DOF_c	e_u, %	e_p, %
1	320	10.069	1.212	1	320	11.279	1.451
2	540	1.112	0.031	2	540	2.943	0.104
4	980	0.253	0.001	4	980	0.579	0.004
8	1860	0.061	0.001	8	1860	0.152	0.001

Table 2. Relative error in the L_2 norm for different numbers of multiscale basis functions. Coefficient k from Test 1 (left) and Test 2 (right) with $C = 10.24$.

M	DOF_c	e_u, %	e_p, %	#iter	M	DOF_c	e_u, %	e_p, %	#iter
1	320	10.334	1.856	9	1	320	11.042	2.155	7
2	540	2.329	0.136	8	2	540	3.585	0.252	6
4	980	2.174	0.129	7	4	980	2.897	0.199	5
8	1860	2.054	0.102	6	8	1860	2.856	0.195	5
Iterations on the fine grid = 5					Iterations on the fine grid = 5				

Table 3. Relative error in the L_2 norm for different numbers of multiscale basis functions. Coefficient k from Test 1 (left) and Test 2 (right) with $C = 34.93$.

M	DOF_c	e_u, %	e_p, %	#iter	M	DOF_c	e_u, %	e_p, %	#iter
1	320	10.389	2.052	24	1	320	11.021	2.374	22
2	540	2.766	0.171	23	2	540	4.059	0.319	20
4	980	2.567	0.151	21	4	980	3.469	0.248	19
8	1860	2.561	0.151	19	8	1860	3.446	0.244	17
Iterations on the fine grid = 21					Iterations on the fine grid = 21				

Table 4. Relative error in the L_2 norm for different numbers of multiscale basis functions. Coefficient k from Test 1 (left) and Test 2 (right) with $C = 1581.14$.

M	DOF_c	e_u, %	e_p, %	#iter	M	DOF_c	e_u, %	e_p, %	#iter
1	320	10.415	2.181	387	1	320	11.033	2.518	399
2	540	2.998	0.197	403	2	540	4.364	0.364	388
4	980	2.814	0.177	389	4	980	3.868	0.294	371
8	1860	2.799	0.175	368	8	1860	3.843	0.291	348
Iterations on the fine grid = 1162					Iterations on the fine grid = 1143				

Table 5. Relative error in the L_2 norm for different numbers of multiscale basis functions. Coefficient k from Test 1 (left) and Test 2 (right) with $C = 71{,}554.17$.

M	DOF_c	e_u, %	e_p, %	#iter	M	DOF_c	e_u, %	e_p, %	#iter
1	320	10.416	2.185	744	1	320	11.033	2.522	794
2	540	3.003	0.198	854	2	540	4.372	0.365	809
4	980	2.821	0.177	886	4	980	3.878	0.295	805
8	1860	2.805	0.176	864	8	1860	3.853	0.291	777
Iterations on the fine grid = 16,489					Iterations on the fine grid = 14,466				

6. Conclusions

In this paper, we conducted a numerical study of the solution of the Darcy-Forchheimer model in high contrast heterogeneous media. To solve this problem, we used the mixed multiscale finite element method. The method showed good accuracy in two model problems. The obtained solutions were compared with the fine grid solution using the mixed finite element method. The study showed that the accuracy of this method depended on the number of multiscale basis functions and was almost independent of the influence of the nonlinear part of the equation. Mixed GMsFEM provided a good solution for any β values, and the accuracy of the method was improved by using more basis functions. The method showed good efficiency, since the number of Picard iterations with a large influence of the nonlinear part was much less than when solving a problem on a fine grid.

Author Contributions: The authors have contributed equally to the work.

Funding: D.S.'s and M.V.'s work was supported by the grant of the Russian Scientific Found N17-71-20055 and the mega-grant of the Russian Federation Government (N 14.Y26.31.0013). E.T.C.'s work was partially supported by the Hong Kong RGC General Research Fund (Projects 14304217 and 14302018) and the CUHK Direct Grant for Research 2018-19. The work of J.H. was supported by the China Postdoctoral Science Foundation Funded Project Grant No. BX20180266, 2018M642991, in part by the National Natural Science Foundation of China Grant No. 11901497, and in part by the Natural Science Foundation of Hunan Province Grant No. 2019JJ50607. The work of Y.H. was supported by the National Natural Science Foundation of China Grant No. 11971410 and in part by the Project of Scientific Research Fund of Hunan Provincial Science and Technology Department (2018WK4006).

Conflicts of Interest: The authors declare no conflict of interest.

References

1. Girault, V.; Wheeler, M.F. Numerical discretization of a Darcy-Forchheimer model. *Numer. Math.* **2008**, *110*, 161–198. [CrossRef]
2. Park, E.J. Mixed finite element methods for generalized Forchheimer flow in porous media. *Numer. Methods Partial Differ. Equ.* **2005**, *21*, 213–228. [CrossRef]
3. Pan, H.; Rui, H. Mixed element method for two-dimensional Darcy-Forchheimer model. *J. Sci. Comput.* **2012**, *52*, 563–587. [CrossRef]
4. Rui, H.; Pan, H. A Block-Centered Finite Difference Method for the Darcy-Forchheimer Model. *SIAM J. Numer. Anal.* **2012**, *50*, 2612–2631. [CrossRef]
5. Rui, H.; Zhao, D.; Pan, H. A block-centered finite difference method for Darcy-Forchheimer model with variable Forchheimer number. *Numer. Methods Partial Differ. Equ.* **2015**, *31*, 1603–1622. [CrossRef]
6. Rui, H.; Liu, W. A Two-Grid Block-Centered Finite Difference Method For Darcy-Forchheimer Flow in Porous Media. *SIAM J. Numer. Anal.* **2015**, *53*, 1941–1962. [CrossRef]
7. Huang, J.; Chen, L.; Rui, H. Multigrid methods for a mixed finite element method of the Darcy-Forchheimer model. *J. Sci. Comput.* **2018**, *74*, 396–411. [CrossRef] [PubMed]
8. Chen, Z.; Hou, T. A mixed multiscale finite element method for elliptic problems with oscillating coefficients. *Math. Comput.* **2003**, *72*, 541–576. [CrossRef]
9. Aarnes, J.E.; Efendiev, Y.; Jiang, L. Mixed multiscale finite element methods using limited global information. *Multiscale Model. Simul.* **2008**, *7*, 655–676. [CrossRef]
10. Aarnes, J.E. On the use of a mixed multiscale finite element method for greaterflexibility and increased speed or improved accuracy in reservoir simulation. *Multiscale Model. Simul.* **2004**, *2*, 421–439. [CrossRef]
11. Aarnes, J.E.; Kippe, V.; Lie, K.A. Mixed multiscale finite elements and streamline methods for reservoir simulation of large geomodels. *Adv. Water Resour.* **2005**, *28*, 257–271. [CrossRef]
12. Chung, E.T.; Efendiev, Y.; Lee, C. Mixed generalized multiscale finite element methods and applications. *Multiscale Model. Simul.* **2015**, *13*, 338–366. [CrossRef]
13. Chung, E.T.; Leung, W.T.; Vasilyeva, M. Mixed GMsFEM for second order elliptic problem in perforated domains. *J. Comput. Appl. Math.* **2016**, *304*, 84–99. [CrossRef]
14. Chung, E.T.; Leung, W.; Vasilyeva, M.; Wang, Y. Multiscale model reduction for transport and flow problems in perforated domains. *J. Comput. Appl. Math.* **2018**, *330*, 519–535. [CrossRef]
15. Zhang, J.; Xing, H. Numerical modeling of non-Darcy flow in near-well region of a geothermal reservoir. *Geothermics* **2012**, *42*, 78–86. [CrossRef]
16. Chang, J.; Nakshatrala, K.B.; Reddy, J.N. Modification to Darcy-Forchheimer model due to pressure-dependent viscosity: Consequences and numerical solutions. *J. Porous Media* **2017**, *20*. [CrossRef]
17. Ciarlet, P.G. *The Finite Element Method for Elliptic Problems*; SIAM: Philadelphia, PA, USA, 2002.
18. Adams, R.; Fournier, J. *Sobolev Spaces*; Academic Press: New York, NY, USA; London, UK; Toronto, ON, Canada, 1975.
19. Chan, H.Y.; Chung, E.; Efendiev, Y. Adaptive mixed GMsFEM for flows in heterogeneous media. *Numer. Math. Theory Methods Appl.* **2016**, *9*, 497–527. [CrossRef]

20. Li, D.; Engler, T.W. Literature review on correlations of the non-Darcy coefficient. In Proceedings of the SPE Permian Basin Oil and Gas Recovery Conference, Midland, TX, YSA, 15–17 May 2001.
21. Muljadi, B.P.; Blunt, M.J.; Raeini, A.Q.; Bijeljic, B. The impact of porous media heterogeneity on non-Darcy flow behaviour from pore-scale simulation. *Adv. Water Resour.* **2016**, *95*, 329–340. [CrossRef]

© 2019 by the authors. Licensee MDPI, Basel, Switzerland. This article is an open access article distributed under the terms and conditions of the Creative Commons Attribution (CC BY) license (http://creativecommons.org/licenses/by/4.0/).

Article

Scattered Data Interpolation and Approximation with Truncated Exponential Radial Basis Function

Qiuyan Xu and Zhiyong Liu *

School of Mathematics and Statistics, Ningxia University, Yinchuan 750021, China; qiuyanxu@nxu.edu.cn
* Correspondence: zhiyong@nxu.edu.cn

Received: 14 October 2019; Accepted: 11 November 2019; Published: 14 November 2019

Abstract: Surface modeling is closely related to interpolation and approximation by using level set methods, radial basis functions methods, and moving least squares methods. Although radial basis functions with global support have a very good approximation effect, this is often accompanied by an ill-conditioned algebraic system. The exceedingly large condition number of the discrete matrix makes the numerical calculation time consuming. The paper introduces a truncated exponential function, which is radial on arbitrary n-dimensional space $\mathbb{R}^n$ and has compact support. The truncated exponential radial function is proven strictly positive definite on $\mathbb{R}^n$ while internal parameter l satisfies $l \geq \lfloor \frac{n}{2} \rfloor + 1$. The error estimates for scattered data interpolation are obtained via the native space approach. To confirm the efficiency of the truncated exponential radial function approximation, the single level interpolation and multilevel interpolation are used for surface modeling, respectively.

Keywords: radial basis functions; native spaces; truncated function; interpolation; approximation; surface modeling

1. Introduction

Radial basis functions can be used to construct trial spaces that have high precision in arbitrary dimensions with arbitrary smoothness. The applications of RBFs (or so-called meshfree methods) can be found in many different areas of science and engineering, including geometric modeling with surfaces [1].The globally supported radial basis functions such as Gaussians or generalized (inverse) multiquadrics have excellent approximation properties. However, they often produce dense discrete systems, which tend to have poor conditioning and lead to a high computational cost. The radial basis functions with compact supports can lead to a very well conditioned sparse system. The goal of this work is to design a truncated exponential function that has compact support and is strictly positive definite and radial on arbitrary n-dimensional space $\mathbb{R}^n$ and to show the advantages of the truncated exponential radial function approximation for surface modeling.

2. Auxiliary Tools

In order to make the paper self-contained and have a complete basis for the theoretical analysis in the later sections, we introduce some concepts and theorems related to radial functions in this section.

2.1. Radial Basis Functions

Definition 1. *A multivariate function* $\Phi : \mathbb{R}^n \to \mathbb{R}$ *is called radial if its value at each point depends only on the distance between that point and the origin, or equivalently provided there exists a univariate function* $\varphi : [0, \infty) \to \mathbb{R}$ *such that* $\Phi(x) = \varphi(r)$ *with* $r = \|x\|$. *Here,* $\|\cdot\|$ *is usually the Euclidean norm. Then, the radial basis functions are defined by translation* $\Phi_k(x) = \varphi(\|x - x_k\|)$ *for any fixed center* $x_k \in \mathbb{R}^n$.

Definition 2. A real-valued continuous function $\Phi : \mathbb{R}^n \to \mathbb{R}$ is called positive definite on $\mathbb{R}^n$ if it is even and:

$$\sum_{j=1}^{N}\sum_{k=1}^{N} c_j c_k \Phi(x_j - x_k) \geq 0 \qquad (1)$$

for any N pairwise different points $x_1, \cdots, x_N \in \mathbb{R}^n$, and $c = [c_1, \cdots, c_N]^T \in \mathbb{R}^N$. It is the Fourier transform of a (positive) measure. The function Φ is strictly positive definite on $\mathbb{R}^n$ if the quadratic (1) is zero only for $c \equiv 0$.

The strictly positive definiteness of the radial function can be characterized by considering the Fourier transform of a univariate function. This is described in the following theorem. Its proof can be found in [2].

Theorem 1. A continuous function $\varphi : [0, \infty) \to \mathbb{R}$ such that $r \to r^{n-1}\varphi(r) \in L^1[0, \infty)$ is strictly positive definite and radial on $\mathbb{R}^n$ if and only if the n-dimensional Fourier transform:

$$\mathcal{F}_n\varphi(r) = \frac{1}{\sqrt{r^{n-2}}} \int_0^\infty \varphi(t) t^{\frac{n}{2}} J_{(n-2)/2}(rt) dt \qquad (2)$$

is non-negative and not identically equal to zero. Here, $J_{(n-2)/2}$ is the classical Bessel function of the first kind of order $(n-2)/2$.

2.2. Multiply Monotonicity

Since Fourier transforms are not always easy to compute, it is convenient to decide whether a function is strictly positive definite and radial on $\mathbb{R}^n$ by the multiply monotonicity for limited choices of n.

Definition 3. A function $\varphi : (0, \infty) \to \mathbb{R}$, which is in $C^{k-2}(0, \infty)$, $k \geq 2$, and for which $(-1)^l \varphi^{(l)}(r)$ is non-negative, non-increasing, and convex for $l = 0, 1, \cdots, k-2$, is called k-times monotone on $(0, \infty)$. In the case $k = 1$, we only require $\varphi \in C(0, \infty)$ to be non-negative and non-increasing.

This definition can be found in the monographs [2,3]. The following Micchelli theorem (see [4]) provides a multiply monotonicity characterization of strictly positive definite radial functions.

Theorem 2. Let $k = \lfloor \frac{n}{2} \rfloor + 2$ be a positive integer. If $\varphi : [0, \infty) \to \mathbb{R}$, $\varphi \in C[0, \infty)$, is k-times monotone on $(0, \infty)$, but not constant, then φ is strictly positive definite and radial on $\mathbb{R}^n$ for any n such that $\lfloor \frac{n}{2} \rfloor \leq k - 2$.

2.3. Native Spaces

Every strictly positive definite function can indeed be associated with a reproducing kernel Hilbert space (or its native space see [5]).

Definition 4. Suppose $\Phi \in C(\mathbb{R}^n) \cap L^1(\mathbb{R}^n)$ is a real-valued strictly positive definite function. Then, the native space of Φ is defined by

$$\mathcal{N}_\Phi(\mathbb{R}^n) = \{f \in L^2(\mathbb{R}^n) \cap C(\mathbb{R}^n) : \frac{\hat{f}}{\sqrt{\hat{\Phi}}} \in L^2(\mathbb{R}^n)\},$$

and equip this space with the norm

$$\|f\|^2_{\mathcal{N}_\Phi(\mathbb{R}^n)} = \int_{\mathbb{R}^n} \frac{|\hat{f}(\omega)|^2}{\hat{\Phi}(\omega)} d\omega < \infty. \qquad (3)$$

For any domain $\Omega \subseteq \mathbb{R}^n$, $\mathcal{N}_\Phi(\Omega)$ is in fact the completion of the pre-Hilbert space $H_\Phi(\Omega) =$ span$\{\Phi(\cdot, \mathbf{y}) : \mathbf{y} \in \Omega\}$. Of course, $\mathcal{N}_\Phi(\Omega)$ contains all functions of the form:

$$f = \sum_{j=1}^{N} c_j \Phi(\cdot, \mathbf{x}_j)$$

provided $\mathbf{x}_j \in \Omega$, and can be assembled with an equivalent norm:

$$\|f\|_{\mathcal{N}_\Phi(\Omega)}^2 = \sum_{j=1}^{N} \sum_{k=1}^{N} c_j c_k \Phi(\mathbf{x}_j, \mathbf{x}_k). \tag{4}$$

Here, $N = \infty$ is also allowed.

3. Truncated Exponential Function

In this section, we design a truncated exponential function:

$$\varphi(r) = (e^{1-r} - 1)_+^l \tag{5}$$

with $r \in \mathbb{R}$, and l is a positive integer. By Definition 1, it becomes apparent that $\Phi(\mathbf{x}) = \varphi(r)$ is a radial function centered on the origin on $\mathbb{R}^n$ when $r = \|\mathbf{x}\|$ and $\mathbf{x} \in \mathbb{R}^n$.

The following theorem characterizes the strictly positive definiteness of $\Phi(\mathbf{x})$.

Theorem 3. *The function $\Phi(\mathbf{x}) = (e^{1-\|\mathbf{x}\|} - 1)_+^l$ is strictly positive definite and radial on $\mathbb{R}^n$ provided parameter l satisfies $l \geq \lfloor \frac{n}{2} \rfloor + 1$.*

Proof. Theorem 2 shows that multiply monotone functions give rise to positive definite radial functions. Therefore, we only need to verify the multiply monotonicity of univariate function $\varphi(r)$ defined by (5).

Obviously, the truncated exponential function $\varphi(r)$ is in $C^{l-1}(0, \infty)$ when $r \in (0, \infty)$ and

$$\varphi(r) = (e^{1-r} - 1)_+^l \geq 0,$$

$$\varphi'(r) = -l e^{1-r} (e^{1-r} - 1)_+^{l-1} \leq 0,$$

$$\varphi''(r) = l(l-1)(e^{1-r})^2 (e^{1-r} - 1)_+^{l-2} + l e^{1-r} (e^{1-r} - 1)_+^{l-1} \geq 0.$$

For any positive integers p and q, $(e^{1-r})^p$ and $(e^{1-r} - 1)_+^q$ are non-negative, but with negative derivatives. Therefore,

$$(-1)^n \varphi^{(n)}(r) \geq 0, \quad n = 0, 1, \cdots, l-1,$$

and $\varphi(r)$ is $(l+1)$-times monotone on $(0, \infty)$. Then, the conclusion follows directly by Theorem 2. □

There are two ways to scale $\varphi(r)$:

(1) In order to make $\varphi(0) = 1$, we can multiply (5) by the positive constant $\frac{1}{(e-1)^l}$ such that $\varphi(r) = \frac{1}{(e-1)^l}(e^{1-r} - 1)_+^l$. Here, $\varphi(r)$ is still strictly positive definite and has the same support as (5).

(2) Adding a shape parameter $\varepsilon > 0$, the scaled truncated exponential function can be given by:

$$\varphi(r) = (e^{1-\varepsilon r} - 1)_+^l. \tag{6}$$

Obviously, a smaller ε causes the function to become flatter and the support to become larger, while increasing ε leads to a more peaked $\varphi(r)$ and therefore localizes its support.

4. Errors in Native Spaces

This section discusses the scattered data interpolation with compactly supported radial basis functions $\Phi(\mathbf{x}, \mathbf{x}_k) = (e^{1-\|\mathbf{x}-\mathbf{x}_k\|} - 1)_+^l$, $\mathbf{x}, \mathbf{x}_k \in \mathbb{R}^n$.

Given a distinct scattered point set $\mathcal{X} = \{\mathbf{x}_1, \mathbf{x}_2, \cdots, \mathbf{x}_N\} \subset \mathbb{R}^n$, the interpolant of target function f can be represented as:

$$P_f(\mathbf{x}) = \sum_{j=1}^{N} c_j \Phi(\mathbf{x}, \mathbf{x}_j), \quad \mathbf{x} \in \mathbb{R}^n. \tag{7}$$

Solving the interpolation problem leads to the following system of linear equations:

$$A\mathbf{c} = \mathbf{y}, \tag{8}$$

where the entries of matrix A are given by $A_{i,j} = \Phi(\mathbf{x}_i, \mathbf{x}_j)$, $i,j = 1, \cdots, N$, $\mathbf{c} = [c_1, \cdots, c_N]^T$, and $\mathbf{y} = [f(\mathbf{x}_1), \cdots, f(\mathbf{x}_N)]^T$. A solution to the system (8) exists and is unique, since the matrix A is positive definite.

Let $\mathbf{u}^*(\mathbf{x}) = [u_1^*(\mathbf{x}), \cdots, u_N^*(\mathbf{x})]^T$ be a cardinal basis vector function, then P_f also has the following form (see [6]):

$$P_f(\mathbf{x}) = \sum_{j=1}^{N} f(\mathbf{x}_j) u_j^*(\mathbf{x}), \quad \mathbf{x} \in \mathbb{R}^n. \tag{9}$$

Comparing (9) with (7), we have:

$$A\mathbf{u}^*(\mathbf{x}) = \mathbf{b}(\mathbf{x}), \tag{10}$$

where $\mathbf{b}(\mathbf{x}) = [\Phi(\mathbf{x}, \mathbf{x}_1), \cdots, \Phi(\mathbf{x}, \mathbf{x}_N)]^T$.

Equation (10) shows a connection between the radial basis functions and the cardinal basis functions.

First, the generic error estimate is as follows.

Theorem 4. *Let $\Omega \subseteq \mathbb{R}^n$, $\mathcal{X} = \{\mathbf{x}_1, \mathbf{x}_2, \cdots, \mathbf{x}_N\} \subset \Omega$ be distinct and $\Phi \in C(\Omega \times \Omega)$ be the truncated exponential radial basis function with $l \geq \lfloor \frac{n}{2} \rfloor + 1$. Denote the interpolant to $f \in \mathcal{N}_\Phi(\Omega)$ on the set $\mathcal{X}$ by P_f. Then, for every $\mathbf{x} \in \Omega$, we have*

$$|f(\mathbf{x}) - P_f(\mathbf{x})| \leq P_{\Phi,\mathcal{X}}(\mathbf{x}) \|f\|_{\mathcal{N}_\Phi(\Omega)}.$$

Here

$$P_{\Phi,\mathcal{X}}(\mathbf{x}) = \sqrt{C - (\mathbf{b}(\mathbf{x}))^T A^{-1} \mathbf{b}(\mathbf{x})}, \quad C = (e-1)^l.$$

Proof. Since $f \in \mathcal{N}_\Phi(\Omega)$, the reproducing property yields

$$f(\mathbf{x}) = \langle f, \Phi(\cdot, \mathbf{x}) \rangle_{\mathcal{N}_\phi(\Omega)}.$$

Then

$$P_f(\mathbf{x}) = \sum_{j=1}^{N} f(\mathbf{x}_j) u_j^*(\mathbf{x}) = \langle f, \sum_{j=1}^{N} u_j^*(\mathbf{x}) \Phi(\cdot, \mathbf{x}_j) \rangle_{\mathcal{N}_\phi(\Omega)}.$$

Applying the Cauchy–Schwarz inequality, we have

$$\begin{aligned} |f(\mathbf{x}) - P_f(\mathbf{x})| &= \left| \langle f, \Phi(\cdot, \mathbf{x}) - \sum_{j=1}^{N} u_j^*(\mathbf{x}) \Phi(\cdot, \mathbf{x}_j) \rangle_{\mathcal{N}_\Phi(\Omega)} \right| \\ &\leq \|f\|_{\mathcal{N}_\Phi(\Omega)} \left\| \Phi(\cdot, \mathbf{x}) - \sum_{j=1}^{N} u_j^*(\mathbf{x}) \Phi(\cdot, \mathbf{x}_j) \right\|_{\mathcal{N}_\Phi(\Omega)}. \end{aligned}$$

Denote the second term as

$$P_{\Phi,\mathcal{X}}(\mathbf{x}) = \left\| \Phi(\cdot,\mathbf{x}) - \sum_{j=1}^{N} u_j^*(\mathbf{x})\Phi(\cdot,\mathbf{x}_j) \right\|_{\mathcal{N}_\Phi(\Omega)}.$$

By the definition of the native space norm and Equation (10), $P_{\Phi,\mathcal{X}}(\mathbf{x})$ can be rewritten as

$$P_{\Phi,\mathcal{X}}(\mathbf{x}) = \sqrt{\Phi(\mathbf{x},\mathbf{x}) - (\mathbf{b}(\mathbf{x}))^T A^{-1} \mathbf{b}(\mathbf{x})}.$$

Then, the conclusion follows directly by the strict positive definiteness of Φ. □

One of the main benefits of Theorem 4 is that we are now able to estimate the interpolation error by computing $P_{\Phi,\mathcal{X}}(\mathbf{x})$. In addition, $P_{\Phi,\mathcal{X}}(\mathbf{x})$ can be used as an indicator for choosing a good shape parameter.

When equipping the dataset $\mathcal{X}$ with a fill distance (or sample density, see [7]):

$$h_{\mathcal{X},\Omega} = \sup_{x \in \Omega} \min_{x_j \in \mathcal{X}} \|\mathbf{x} - \mathbf{x}_j\|,$$

for any symmetric and strictly positive definite $\Phi \in C^{2k}(\Omega \times \Omega)$, the following generic error estimate can be obtained.

Theorem 5. *Suppose $\Omega \subseteq \mathbb{R}^n$ is bounded and satisfies an interior cone condition. Suppose $\Phi \in C^{2k}(\Omega \times \Omega)$ is symmetric and strictly positive definite. Denote the interpolant to $f \in \mathcal{N}_\Phi(\Omega)$ on the set $\mathcal{X}$ by P_f. Then, there exist some positive constants h_0 and C such that:*

$$|f(x) - P_f(x)| \le C h_{\mathcal{X},\Omega}^k \sqrt{D_\Phi(x)} \|f\|_{\mathcal{N}_\Phi(\Omega)},$$

provided $h_{\mathcal{X},\Omega} \le h_0$. Here

$$D_\Phi(x) = \max_{|\beta|=2k} \max_{w,z \in \Omega \cap B(x,ch_{\mathcal{X},\Omega})} |D_2^\beta \Phi(w,z)|$$

with $B(x, ch_{\mathcal{X},\Omega})$ denoting the ball of radius $ch_{\mathcal{X},\Omega}$ centered at x.

Proof. The estimate can be obtained by applying the Taylor expansion. The technical details can be found in [2,3]. □

Since the truncated radial basis function Φ is only in $C^0(\Omega \times \Omega)$, $h_{\mathcal{X},\Omega}^k$ is vanishing in the above error estimate of Theorem 5. Therefore, we need to bound the $D_\Phi(x)$ by some additional powers of $h_{\mathcal{X},\Omega}$ in order to obtain the estimate in terms of fill distance. The resulting theorem is as follows.

Theorem 6. *Suppose $\Omega \subseteq \mathbb{R}^n$ is bounded and satisfies an interior cone condition. Suppose Φ is the truncated exponential radial basis function with $l \ge \lfloor \frac{n}{2} \rfloor + 1$. Denote the interpolant to $f \in \mathcal{N}_\Phi(\Omega)$ on the set $\mathcal{X}$ by P_f. Then, there exist some positive constants h_0 and C such that:*

$$|f(x) - P_f(x)| \le C h_{\mathcal{X},\Omega}^{\frac{1}{2}} \|f\|_{\mathcal{N}_\Phi(\Omega)},$$

provided $h_{\mathcal{X},\Omega} \le h_0$.

Proof. From [2], for C^0 functions, the factor $D_\Phi(\mathbf{x})$ can be expressed as:

$$D_\Phi(\mathbf{x}) = \|\Phi\|_{L^\infty(B(0, 2ch_{\chi,\Omega}))}$$

independent of $\mathbf{x}$. Selecting $h_0 \leq \frac{1}{4c}$, we bound the $D_\Phi(\mathbf{x})$ determined by the truncated exponential radial basis function.

Using the definition of Φ and Lagrange's mean value theorem, we have:

$$\begin{aligned}
\|\Phi\|_{L^\infty(B(0, 2ch_{\chi,\Omega}))} &= \max_{r \in (0, 4ch_{\chi,\Omega})} |e^{1-r} - 1|^l \\
&\leq C \max_{r \in (0, 4ch_{\chi,\Omega})} |1 - r|^l \\
&= C \|\Psi\|_{L^\infty(B(0, 2ch_{\chi,\Omega}))}
\end{aligned}$$

with Ψ denoting the truncated power radial basis function. From [2],

$$\|\Psi\|_{L^\infty(B(0, 2ch_{\chi,\Omega}))} \leq C h_{\chi,\Omega}^{\frac{1}{2}}.$$

□

5. Numerical Experiments

5.1. Single-Level Approximation

This subsection shows how our truncated exponential radial basis function (TERBF) works at a single level. Our first 2D target surface is the standard Franke's function. In the experiments, we let the kernel Φ in (7) be the truncated exponential radial function $\Phi(\mathbf{x}) = (e^{1-\epsilon\|\mathbf{x}\|} - 1)_+^2$. A Halton point set with increasingly greater data density is generated in domain $[0,1]^2$. Tables 1–8 list the test results of Gaussian interpolation, MQ (Multiquadrics) interpolation, IMQ (Inverse Multiquadrics) interpolation, and TERBF interpolation with different values of ϵ respectively. In the tables, the RMS-error is computed by

$$\text{RMS-error} = \sqrt{\frac{1}{M}\sum_{k=1}^{M}[f(\xi_k) - P_f(\xi_k)]^2} = \frac{1}{\sqrt{M}}\|f - P_f\|_2,$$

where ξ_k are the evaluation points. The rate listed in the Tables is computed using the formula:

$$\text{rate}_k = \frac{\ln(e_{k-1}/e_k)}{\ln(h_{k-1}/h_k)}, \quad k = 2, 3, 4, 5, 6,$$

where e_k is the RMS-error for experiment number k and h_k is the fill distance of the k-level. cond(A) is the condition number of the interpolation matrix defined by (8). From Tables 1–6, we observe that the globally supported radial basis functions (Gaussian, MQ, IMQ) can obtain ideal accuracy when assembling a smaller value of ϵ. However, the condition number of the interpolation matrix will become surprisingly large as the scattered data increase. We note that MATLAB issues a "matrix close to singular" warning when carrying out Gaussian and MQ interpolation experiments for $N = 1089, 4225$ and $\epsilon = 10$. Tables 7 and 8 show that TERBF interpolation can not only keep better approximation accuracy, but also produce a well conditioned interpolation matrix. Even for $N = 4225$ and $\epsilon = 0.7$, the condition number of the presented method is relatively smaller (about 10^5). The change of RMS-error with varying ϵ values is displayed in Figure 1. We see that the error curves of Gaussian and MQ interpolation are not monotonic and even become erratic for the largest datasets. However, the curves of IMQ and TERBF interpolation are relatively smooth. In particular, TERBF greatly improves the condition number of the interpolation matrix. To show the application of TERBF approximation to compact 3D images, we interpolate Beethoven data in Figure 2 and Stanford bunny in Figure 3. Numerical experiments suggest that TERBF interpolation is essentially faster than the scattered data interpolation with globally supported radial basis functions. However, we observe that TERBF interpolation causes some artifacts such as the extra surface fragment near the bunny's ear from the left part of Figure 3. This is because the interpolating implicit surface has a narrow band support. It will be better if the sample density is smaller than the width of the support band (see the right part of Figure 3). Similar observations have been reported in Fasshauer's book [3], where a partition of unity fits based on Wendland's C^2 function was used. The same observation was also made in [1].

Table 1. Gaussian interpolation to the 2D Franke's function with $\varepsilon = 20$.

N	RMS-Error	Rate	cond(A)
9	3.633326×10^{-1}	-	$1.000028 \times 10^{+0}$
25	3.138226×10^{-1}	0.211341	$1.006645 \times 10^{+0}$
81	2.003929×10^{-1}	0.647118	$3.170400 \times 10^{+0}$
289	6.616318×10^{-2}	1.598731	$3.761572 \times 10^{+1}$
1089	1.205109×10^{-2}	2.456865	$1.925205 \times 10^{+5}$
4225	2.908614×10^{-4}	5.372688	$2.687885 \times 10^{+16}$

Table 2. Gaussian interpolation to the 2D Franke's function with $\varepsilon = 10$.

N	RMS-Error	Rate	cond(A)
9	3.256546×10^{-1}	-	$1.129919 \times 10^{+0}$
25	1.722746×10^{-1}	0.918633	$1.667637 \times 10^{+0}$
81	5.465624×10^{-2}	1.656252	$2.601726 \times 10^{+1}$
289	1.391350×10^{-2}	1.973901	$7.316820 \times 10^{+4}$
1089	3.273510×10^{-4}	5.409503	$1.179104 \times 10^{+16}$
4225	1.135157×10^{-6}	8.171803	$1.906108 \times 10^{+20}$

Table 3. MQ interpolation to the 2D Franke's function with $\varepsilon = 20$.

N	RMS-Error	Rate	cond(A)
9	1.224583×10^{-1}	-	$5.366051 \times 10^{+1}$
25	5.646454×10^{-2}	1.116874	$3.124063 \times 10^{+2}$
81	6.998841×10^{-3}	3.012157	$5.534539 \times 10^{+3}$
289	1.418117×10^{-3}	2.303139	$2.324743 \times 10^{+5}$
1089	3.627073×10^{-4}	1.967099	$8.803829 \times 10^{+7}$
4225	4.969932×10^{-5}	2.867508	$5.331981 \times 10^{+11}$

Table 4. MQ interpolation to the 2D Franke's function with $\varepsilon = 10$.

N	RMS-Error	Rate	cond(A)
9	1.146184×10^{-1}	-	$8.464360 \times 10^{+1}$
25	5.193997×10^{-2}	1.141921	$6.680998 \times 10^{+2}$
81	4.534144×10^{-3}	3.517943	$2.158362 \times 10^{+4}$
289	9.608696×10^{-4}	2.238418	$5.033541 \times 10^{+6}$
1089	1.506154×10^{-4}	2.673471	$3.025049 \times 10^{+10}$
4225	4.603113×10^{-6}	5.032116	$5.613893 \times 10^{+16}$

Table 5. IMQ interpolation to the 2D Franke's function with $\varepsilon = 20$.

N	RMS-Error	Rate	cond(A)
9	2.491443×10^{-1}	-	$2.733942 \times 10^{+0}$
25	9.914856×10^{-2}	1.329318	$6.933813 \times 10^{+0}$
81	3.257319×10^{-2}	1.605907	$5.444834 \times 10^{+1}$
289	1.159691×10^{-2}	1.489945	$1.022341 \times 10^{+3}$
1089	3.420734×10^{-3}	1.761362	$1.850967 \times 10^{+5}$
4225	6.703871×10^{-4}	2.351240	$5.607685 \times 10^{+8}$

Table 6. IMQ interpolation to the 2D Franke's function with $\varepsilon = 10$.

N	RMS-Error	Rate	cond(A)
9	2.065836×10^{-1}	-	$5.995564 \times 10^{+0}$
25	5.366442×10^{-2}	1.944688	$2.312141 \times 10^{+1}$
81	1.517723×10^{-2}	1.822057	$4.053520 \times 10^{+2}$
289	5.181480×10^{-3}	1.550472	$3.889766 \times 10^{+4}$
1089	9.630601×10^{-4}	2.427667	$1.155244 \times 10^{+8}$
4225	4.615820×10^{-5}	4.382967	$1.158439 \times 10^{+14}$

Table 7. TERBF interpolation to the 2D Franke's function with $\varepsilon = 1$.

N	RMS-Error	Rate	cond(A)
9	1.951235×10^{-1}	-	$6.639719 \times 10^{+0}$
25	5.018953×10^{-2}	1.958929	$2.405994 \times 10^{+1}$
81	1.628459×10^{-2}	1.623879	$1.669026 \times 10^{+2}$
289	6.727682×10^{-3}	1.275326	$1.250365 \times 10^{+3}$
1089	2.402630×10^{-3}	1.485495	$1.058555 \times 10^{+4}$
4225	9.728457×10^{-4}	1.304332	$9.410946 \times 10^{+4}$

Table 8. TERBF interpolation to the 2D Franke's function with $\varepsilon = 0.7$.

N	RMS-Error	Rate	cond(A)
9	1.728785×10^{-1}	-	$1.275042 \times 10^{+1}$
25	4.535991×10^{-2}	1.930269	$5.066809 \times 10^{+1}$
81	1.335521×10^{-2}	1.764015	$3.608813 \times 10^{+2}$
289	5.013012×10^{-3}	1.413653	$2.719227 \times 10^{+3}$
1089	1.773595×10^{-3}	1.499001	$2.305630 \times 10^{+4}$
4225	7.107796×10^{-4}	1.319203	$2.050036 \times 10^{+5}$

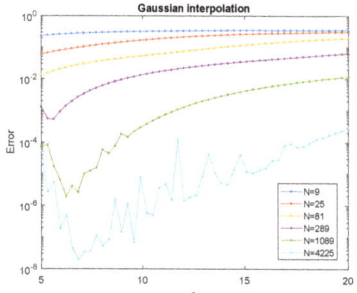

(**a**) ε curves of Gaussian interpolation

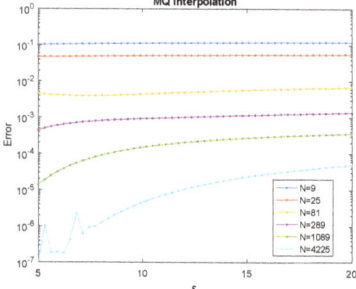

(**b**) ε curves of MQ interpolation

Figure 1. *Cont.*

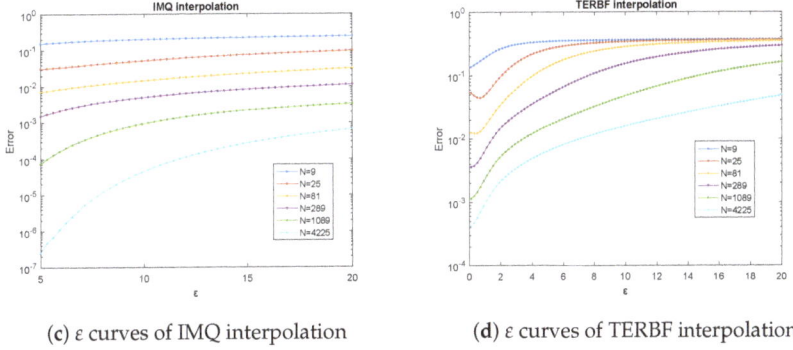

(c) ε curves of IMQ interpolation   (d) ε curves of TERBF interpolation

Figure 1. RMS-error curves for Gaussian, MQ, IMQ, and TERBF interpolations.

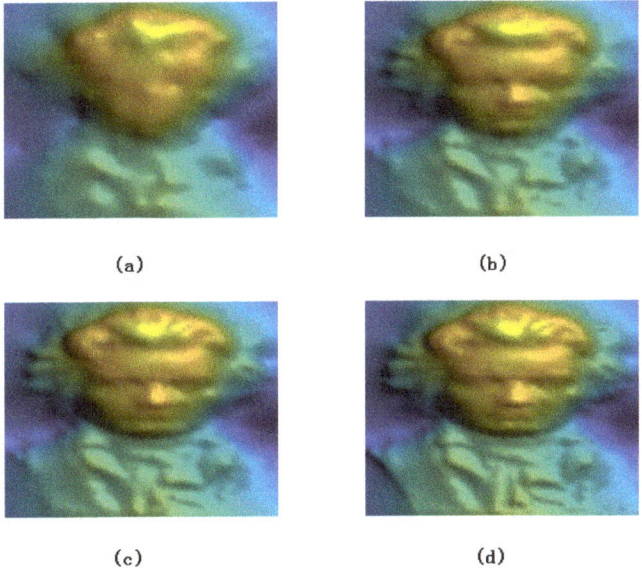

Figure 2. TERBF approximation of the Beethoven data. From top left to bottom right: 163 (**a**), 663 (**b**), 1163 (**c**), and 2663 (**d**) points.

Figure 3. TERBF approximation of the Stanford bunny with 453 (**left**) and 8171 (**right**) data points.

5.2. Multilevel Approximation

The multilevel scattered approximation was implemented first in [8] and then studied by a number of other researchers [9–13]. In the multilevel algorithm, the residual can be formed on the coarsest level first and then be approximated on the later finer level by the compactly supported radial basis functions with gradually smaller support. This process can be repeated and be stopped on the finest level. An advantage of this multilevel interpolation algorithm is its recursive property (i.e., the same routine can be applied recursively at each level in the programming language), of course the disadvantage being the allocation that memory needs.

In this experiment, suppose a 3D target surface is an explicit function $f(x,y,z) = 64x(1-x)y(1-y)z(1-z)$. We generate a uniform points set in the domain $[0,1]^3$, with levels $k = 1,2,3,4$ and $N = 27, 125, 729, 4913$. The scale parameter $\varepsilon = 0.07 \times 2^{[0:3]}$, and $l = 3$. The corresponding slice plots, the iso-surfaces, and slice plots of the absolute error are shown in Figures 4–7. Both the iso-surfaces and the slice plots are color coded according to the absolute error. At each level, the trial space is constructed by a series of truncated exponential radial basis functions with varying support radii. Hence, the multilevel approximation algorithm can produce a well conditioned sparse discrete algebraic system in each recursion and keep ideal approximation accuracy at the same time. Numerical experiments show that TERBF multilevel interpolation is very effective for 3D explicit surface approximation. These observations can be found from Figures 4–7. Similar experiments and observations are reported in detail in Fasshauer's book [3], where Wendland's function C^4 has been used for approximation. However, to improve the allocation memory needs of the multilevel algorithm, we can make use of the hierarchical collocation method developed in [13].

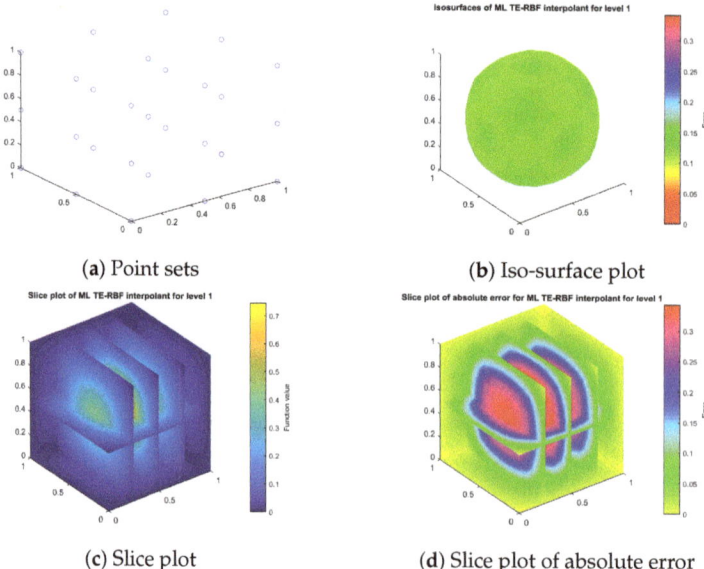

(a) Point sets
(b) Iso-surface plot
(c) Slice plot
(d) Slice plot of absolute error

Figure 4. Fits and errors at Level 1.

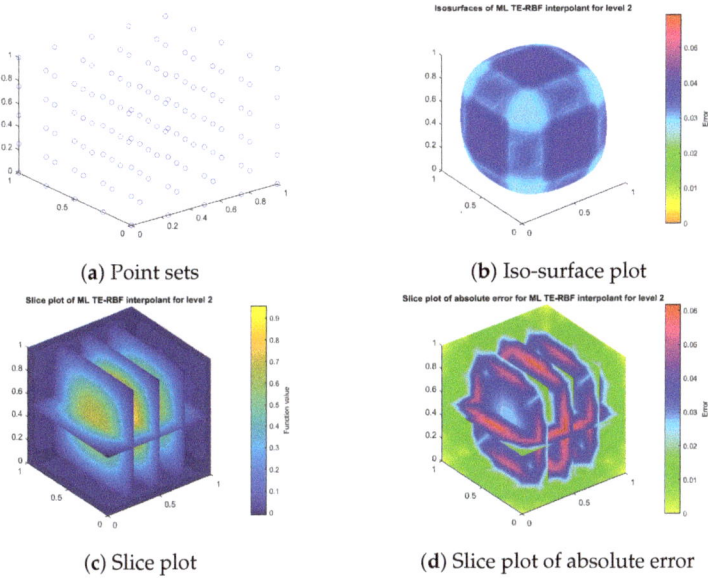

Figure 5. Fits and errors at Level 2.

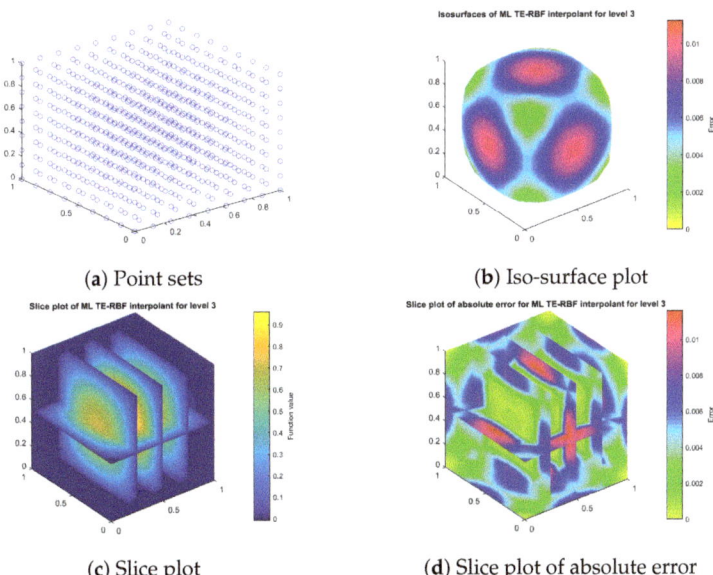

Figure 6. Fits and errors at Level 3.

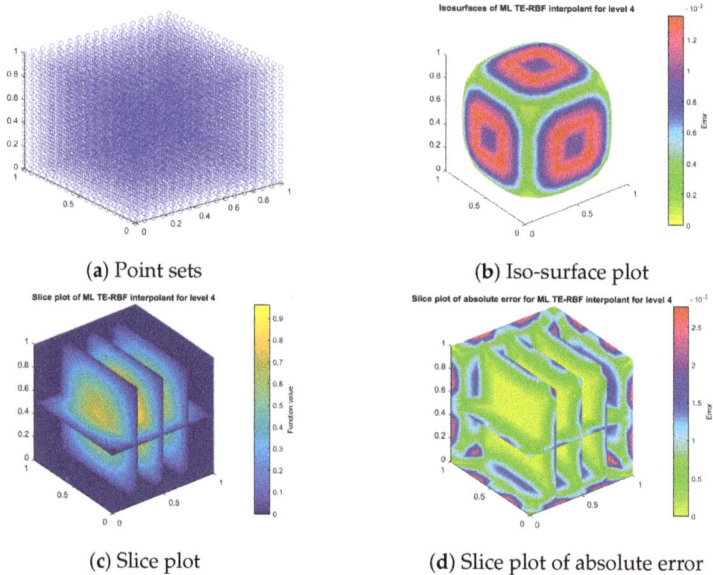

(a) Point sets (b) Iso-surface plot (c) Slice plot (d) Slice plot of absolute error

Figure 7. Fits and errors at Level 4.

6. Conclusions

The truncated exponential radial function, which has compact support, was introduced in the paper. The strictly positive definiteness of TERBF was proven via the multiply monotonicity approach, and the interpolation error estimates were obtained via the native space approach. Moreover, the TERBF was applied to 2D/3D scattered data interpolation and surface modeling successfully.

However, we found that $\Phi(\mathbf{x}) = (e^{1-\epsilon\|\mathbf{x}\|} - 1)_+^l$ was only in C^0 space. In the error estimates in terms of the fill distance, the power of $h_{\mathcal{X},\Omega}$ was only $1/2$. There are many possibilities for enhancement of TERBF approximation:

(1) We can construct new strictly positive definite radial functions with finite smoothness from the given $\Phi(\mathbf{x})$ by a "dimension-walk" technique.

(2) We can do in-depth analysis of the characterization of TERBF in terms of Fourier transforms established by Bochner and Schoenberg's theorems.

(3) TERBF can also be used for the numerical solution of partial differential equations. The convergence proof will depend on the approximation of TERBF trial spaces, the appropriate inverse inequality, and the sampling theorem.

Author Contributions: Conceptualization, Methodology and Writing–original draft preparation, Q.X.; Formal analysis and Writing—review and editing, Z.L.

Funding: The research of the first author was partially supported by the Natural Science Foundations of Ningxia Province (No. NZ2018AAC03026) and the Fourth Batch of the Ningxia Youth Talents Supporting Program (No. TJGC2019012). The research of the second author was partially supported by the Natural Science Foundations of China (No. 11501313), the Natural Science Foundations of Ningxia Province (No. 2019AAC02001), the Project funded by the China Postdoctoral Science Foundation (No. 2017M621343), and the Third Batch of the Ningxia Youth Talents Supporting Program (No. TJGC2018037).

Acknowledgments: The authors would like to thank the Editor and two unknown reviewers who made valuable comments on an earlier version of this paper. The authors used some Halton datasets and drew lessons from partial codes from Fasshauer's book [3]. We are grateful to [3] for its free CD, which contains many MATLAB codes.

Conflicts of Interest: The authors declare no conflict of interest.

References

1. Ohtake, Y.; Belyaev, A.; Seidel, H.P. 3D scattered data interpolation and approximation with multilevel compactly supported RBFs. *Graph. Model.* **2005**, *67*, 150–165. [CrossRef]
2. Wendland, H. *Scattered Data Approximation*; Cambridge University Press: Cambridge, UK, 2005.
3. Fasshauer, G.E. *Meshfree Approximation Methods with MATLAB*; World Scientific Publishers: Singapore, 2007.
4. Micchelli, C.A. Interpolation of scattered data: Distance matrices and conditionally positive definite functions. *Constr. Approx.* **1986**, *2*, 11–22. [CrossRef]
5. Schaback, R. A unified theory of radial basis functions: Native Hilbert spaces for radial basis functions II. *J. Comp. Appl. Math.* **2000**, *121*, 165–177. [CrossRef]
6. De Marchi, S.; Perracchiono, E. *Lectures on Radial Basis Functions*; Department of Mathematics, "Tullio Levi-Civita", University of Padova: Padova, Italy. Available online: https://www.google.com/url?sa=t&rct=j&q=&esrc=s&source=web&cd=1&ved=2ahUKEwjkuuu01ejlAhW9xosBHaZ0Ct8QFjAAegQIABAC&url=https%3A%2F%2Fwww.math.unipd.it%2F~demarchi%2FRBF%2FLectureNotes_new.pdf&usg=AOvVaw0sDK5WcNE1POWoa_lVur9v (accessed on 20 October 2019).
7. Bernard, C.P.; Mallat, S.G.; Slotine, J.J. Scattered data interpolation with wavelet trees. In *Curve and Surface Fitting (Saint-Malo, 2002)*; Nashboro Press: Brentwood, TN, USA, 2003; pp. 59–64.
8. Floater, M.S.; Iske, A. Multistep scattered data interpolation using compactly supported radial basis functions. *J. Comput. Appl. Math.* **1996**, *73*, 65–78. [CrossRef]
9. Chen, C.S.; Ganesh, M.; Golberg, M.A.; Cheng, A.H.D. Multilevel compact radial functions based computational schemes for some elliptic problems. *Comput. Math. Appl.* **2002**, *43*, 359–378. [CrossRef]
10. Chernih, A.; Gia, Q.T.L. Multiscale methods with compactly supported radial basis functions for the Stokes problem on bounded domains. *Adv. Comput. Math.* **2016**, *42*, 1187–1208. [CrossRef]
11. Farrell, P.; Wendland, H. RBF multiscale collocation for second order elliptic boundary value problems. *SIAM J. Numer. Anal.* **2013**, *51*, 2403–2425. [CrossRef]
12. Fasshauer, G.E.; Jerome, J.W. Multistep approximation algorithms: Improved convergence rates through postconditioning with smoothing kernels. *Adv. Comput. Math.* **1999**, *10*, 1–27. [CrossRef]
13. Liu, Z.; Xu, Q. A Multiscale RBF Collocation Method for the Numerical Solution of Partial Differential Equations. *Mathematics* **2019**, *7*, 964. [CrossRef]

© 2019 by the authors. Licensee MDPI, Basel, Switzerland. This article is an open access article distributed under the terms and conditions of the Creative Commons Attribution (CC BY) license (http://creativecommons.org/licenses/by/4.0/).

Article

Universal Function Approximation by Deep Neural Nets with Bounded Width and ReLU Activations

Boris Hanin

Department of Mathematics, Texas A&M, College Station, TX 77843, USA; bhanin@math.tamu.edu

Received: 29 September 2019; Accepted: 16 October 2019; Published: 18 October 2019

Abstract: This article concerns the expressive power of depth in neural nets with ReLU activations and a bounded width. We are particularly interested in the following questions: What is the minimal width $w_{\min}(d)$ so that ReLU nets of width $w_{\min}(d)$ (and arbitrary depth) can approximate any continuous function on the unit cube $[0,1]^d$ arbitrarily well? For ReLU nets near this minimal width, what can one say about the depth necessary to approximate a given function? We obtain an essentially complete answer to these questions for convex functions. Our approach is based on the observation that, due to the convexity of the ReLU activation, ReLU nets are particularly well suited to represent convex functions. In particular, we prove that ReLU nets with width $d+1$ can approximate any continuous convex function of d variables arbitrarily well. These results then give quantitative depth estimates for the rate of approximation of any continuous scalar function on the d-dimensional cube $[0,1]^d$ by ReLU nets with width $d+3$.

Keywords: Deep Neural Nets; ReLU Networks; Approximation Theory

1. Introduction

Over the past several years, neural nets, particularly deep nets, have become the state-of-the-art in a remarkable number of machine learning problems, from mastering go to image recognition/segmentation and machine translation (see the review article [1] for more background). Despite all their practical successes, a robust theory of why they work so well is in its infancy. Much of the work to date has focused on the problem of explaining and quantifying the expressivity (the ability to approximate a rich class of functions) of deep neural nets [2–11]. Expressivity can be seen both as an effect of both depth and width. It has been known since at least the work of Cybenko [12] and Hornik-Stinchcombe-White [13] that if no constraint is placed on the width of a hidden layer, then a single hidden layer is enough to approximate essentially any function. The purpose of this article, in contrast, is to investigate the "effect of depth without the aid of width." More precisely, for each $d \geq 1$, we would like to estimate:

$$w_{\min}(d) := \min\left\{ w \in \mathbb{N} \,\middle|\, \begin{array}{c} \text{ReLU nets of width } w \text{ can approximate any} \\ \text{positive continuous function on } [0,1]^d \text{ arbitrarily well} \end{array} \right\}. \quad (1)$$

Here, $\mathbb{N} = \{0, 1, 2, \ldots\}$ are the natural numbers and ReLU is the so-called "rectified linear unit," $\mathrm{ReLU}(t) = \max\{0, t\}$, which is the most popular non-linearity used in practice (see (4) for the exact definition). In Theorem 1, we prove that $w_{\min}(d) \leq d+2$. This raises two questions:

Q1. Is the estimate in the previous line sharp?

Q2. How efficiently can ReLU nets of a given width $w \geq w_{\min}(d)$ approximate a given continuous function of d variables?

A priori, it is not clear how to estimate $\omega_{min}(d)$ and whether it is even finite. One of the contributions of this article is to provide reasonable bounds on $\omega_{min}(d)$ (see Theorem 1). Moreover, we also provide quantitative estimates on the corresponding rate of approximation. On the subject of Q1, we will prove in forthcoming work with M.Sellke [14] that in fact, $\omega_{min}(d) = d + 1$. When $d = 1$, the lower bound is simple to check, and the upper bound follows for example from Theorem 3.1 in [5]. The main results in this article, however, concern Q1 and Q2 for convex functions. For instance, we prove in Theorem 1 that:

$$w_{min}^{conv}(d) \leq d + 1, \tag{2}$$

where:

$$w_{min}^{conv}(d) := \min\left\{ w \in \mathbb{N} \,\middle|\, \begin{array}{c} \text{ReLU nets of width } w \text{ can approximate any} \\ \text{positive convex function on } [0,1]^d \text{ arbitrarily well} \end{array} \right\}. \tag{3}$$

This illustrates a central point of the present paper: the convexity of the ReLU activation makes ReLU nets well-adapted to representing convex functions on $[0,1]^d$.

Theorem 1 also addresses Q2 by providing quantitative estimates on the depth of a ReLU net with width $d + 1$ that approximates a given convex function. We provide similar depth estimates for arbitrary continuous functions on $[0,1]^d$, but this time for nets of width $d + 3$. Several of our depth estimates are based on the work of Balázs-György-Szepesvári [15] on max-affine estimators in convex regression.

In order to prove Theorem 1, we must understand what functions can be exactly computed by a ReLU net. Such functions are always piecewise affine, and we prove in Theorem 2 the converse: every piecewise affine function on $[0,1]^d$ can be exactly represented by a ReLU net with hidden layer width at most $d + 3$. Moreover, we prove that the depth of the network that computes such a function is bounded by the number affine pieces it contains. This extends the results of Arora-Basu-Mianjy-Mukherjee (e.g., Theorem 2.1 and Corollary 2.2 in [2]).

Convex functions again play a special role. We show that every convex function on $[0,1]^d$ that is piecewise affine with N pieces can be represented exactly by a ReLU net with width $d + 1$ and depth N.

2. Statement of Results

To state our results precisely, we set notation and recall several definitions. For $d \geq 1$ and a continuous function $f : [0,1]^d \to \mathbb{R}$, write:

$$\|f\|_{C^0} := \sup_{x \in [0,1]^d} |f(x)|.$$

Further, denote by:

$$\omega_f(\varepsilon) := \sup\{|f(x) - f(y)| \mid |x - y| \leq \varepsilon\}$$

the modulus of continuity of f, whose value at ε is the maximum that f can change when its argument moves by at most ε. Note that by the definition of a continuous function, $\omega_f(\varepsilon) \to 0$ as $\varepsilon \to 0$. Next, given d_{in}, d_{out}, and $w \geq 1$, we define a feed-forward neural net with ReLU activations, input dimension d_{in}, hidden layer width w, depth n, and output dimension d_{out} to be any member of the finite-dimensional family of functions:

$$\text{ReLU} \circ A_n \circ \cdots \circ \text{ReLU} \circ A_1 \circ \text{ReLU} \circ A_1 \tag{4}$$

that map $\mathbb{R}^d$ to $\mathbb{R}^{d_{out}}_+ = \{x = (x_1, \ldots, x_{d_{out}}) \in \mathbb{R}^{d_{out}} \mid x_i \geq 0\}$. In (4),

$$A_j : \mathbb{R}^w \to \mathbb{R}^w, \ j = 2, \ldots, n-1, \qquad A_1 : \mathbb{R}^{d_{in}} \to \mathbb{R}^w, \ A_n : \mathbb{R}^w \to \mathbb{R}^{d_{out}}$$

are affine transformations, and for every $m \geq 1$:

$$\text{ReLU}(x_1, \ldots, x_m) = (\max\{0, x_1\}, \ldots, \max\{0, x_m\}).$$

We often denote such a net by $\mathcal{N}$ and write:

$$f_{\mathcal{N}}(x) := \text{ReLU} \circ A_n \circ \cdots \circ \text{ReLU} \circ A_1 \circ \text{ReLU} \circ A_1(x)$$

for the function it computes. Our first result contrasts both the width and depth required to approximate continuous, convex, and smooth functions by ReLU nets.

Theorem 1. *Let $d \geq 1$ and $f : [0,1]^d \to \mathbb{R}_+$ be a positive function with $\|f\|_{C^0} = 1$. We have the following three cases:*

1. **(f is continuous)** *There exists a sequence of feed-forward neural nets $\mathcal{N}_k$ with ReLU activations, input dimension d, hidden layer width $d + 2$, and output dimension 1, such that:*

$$\lim_{k \to \infty} \|f - f_{\mathcal{N}_k}\|_{C^0} = 0. \tag{5}$$

In particular, $w_{min}(d) \leq d + 2$. Moreover, write ω_f for the modulus of continuity of f, and fix $\varepsilon > 0$. There exists a feed-forward neural net $\mathcal{N}_\varepsilon$ with ReLU activations, input dimension d, hidden layer width $d + 3$, output dimension 1, and:

$$\text{depth}(\mathcal{N}_\varepsilon) = \frac{2 \cdot d!}{\omega_f(\varepsilon)^d} \tag{6}$$

such that:

$$\|f - f_{\mathcal{N}_\varepsilon}\|_{C^0} \leq \varepsilon. \tag{7}$$

2. **(f is convex)** *There exists a sequence of feed-forward neural nets $\mathcal{N}_k$ with ReLU activations, input dimension d, hidden layer width $d + 1$, and output dimension 1, such that:*

$$\lim_{k \to \infty} \|f - f_{\mathcal{N}_k}\|_{C^0} = 0. \tag{8}$$

Hence, $\omega_{min}^{conv}(d) \leq d + 1$. Further, there exists $C > 0$ such that if f is both convex and Lipschitz with Lipschitz constant L, then the nets $\mathcal{N}_k$ in (8) can be taken to satisfy:

$$\text{depth}(\mathcal{N}_k) = k + 1, \qquad \|f - f_{\mathcal{N}_k}\|_{C^0} \leq CLd^{3/2}k^{-2/d}. \tag{9}$$

3. **(f is smooth)** *There exists a constant K depending only on d and a constant C depending only on the maximum of the first K derivative of f such that for every $k \geq 3$, the width $d + 2$ nets $\mathcal{N}_k$ in (5) can be chosen so that:*

$$\text{depth}(\mathcal{N}_k) = k, \qquad \|f - f_{\mathcal{N}_k}\|_{C^0} \leq C(k-2)^{-1/d}. \tag{10}$$

The main novelty of Theorem 1 is the width estimate $w_{min}^{conv}(d) \leq d + 1$ and the quantitative depth estimates (9) for convex functions, as well as the analogous estimates (6) and (7) for continuous functions. Let us briefly explain the origin of the other estimates. The relation (5) and the corresponding estimate $w_{min}(d) \leq d + 2$ are a combination of the well-known fact that ReLU nets with one hidden layer can approximate any continuous function and a simple procedure by which a ReLU net with input dimension d and a single hidden layer of width n can be replaced by another ReLU net that computes the same function, but has depth $n + 2$ and width $d + 2$. For these width $d + 2$ nets, we are unaware of how to obtain quantitative estimates on the depth required to approximate a fixed continuous function to a given precision. At the expense of changing the width of our ReLU nets from $d + 2$ to $d + 3$, however, we furnish the estimates (6) and (7). On the other hand, using Theorem 3.1 in [5], when f is sufficiently smooth, we obtain the depth estimates (10) for width $d+2$ ReLU nets. Indeed, since we are working on a compact set $[0,1]^d$, the smoothness classes $W_{w,q,\gamma}$ from [5] reduce to classes of functions that have sufficiently many bounded derivatives.

Our next result concerns the exact representation of piecewise affine functions by ReLU nets. Instead of measuring the complexity of such a function by its Lipschitz constant or modulus of continuity, the complexity of a piecewise affine function can be thought of as the minimal number of affine pieces needed to define it.

Theorem 2. *Let $d \geq 1$ and $f : [0,1]^d \to \mathbb{R}_+$ be the function computed by some ReLU net with input dimension d, output dimension 1, and arbitrary width. There exist affine functions $g_\alpha, h_\beta : [0,1]^d \to \mathbb{R}$ such that f can be written as the difference of positive convex functions:*

$$f = g - h, \qquad g := \max_{1 \leq \alpha \leq N} g_\alpha, \qquad h := \max_{1 \leq \beta \leq M} h_\beta. \qquad (11)$$

Moreover, there exists a feed-forward neural net $\mathcal{N}$ with ReLU activations, input dimension d, hidden layer width $d+3$, output dimension 1, and:

$$\mathrm{depth}(\mathcal{N}) = 2(M+N) \qquad (12)$$

that computes f exactly. Finally, if f is convex (and hence, h vanishes), then the width of $\mathcal{N}$ can be taken to be $d+1$, and the depth can be taken to be N.

The fact that the function computed by a ReLU net can be written as (11) follows from Theorem 2.1 in [2]. The novelty in Theorem 2 is therefore the uniform width estimate $d+3$ in the representation on any function computed by a ReLU net and the $d+1$ width estimate for convex functions. Theorem 2 will be used in the proof of Theorem 1.

3. Relation to Previous Work

This article is related to several strands of prior work:

1. Theorems 1 and 2 are "deep and narrow" analogs of the well-known "shallow and wide" universal approximation results (e.g., Cybenko [12] and Hornik-Stinchcombe-White [13]) for feed-forward neural nets. Those articles show that essentially any scalar function $f : [0,1]^d \to \mathbb{R}$ on the d-dimensional unit cube can be arbitrarily well approximated by a feed-forward neural net with a single hidden layer with arbitrary width. Such results hold for a wide class of nonlinear activations, but are not particularly illuminating from the point of understanding the expressive advantages of depth in neural nets.
2. The results in this article complement the work of Liao-Mhaskar-Poggio [3] and Mhaskar-Poggio [5], who considered the advantages of depth for representing certain hierarchical or compositional functions by neural nets with both ReLU and non-ReLU activations. Their results (e.g., Theorem 1 in [3] and Theorem 3.1 in [5]) give bounds on the width for approximation both for shallow and certain deep hierarchical nets.
3. Theorems 1 and 2 are also quantitative analogs of Corollary 2.2 and Theorem 2.4 in the work of Arora-Basu-Mianjy-Mukerjee [2]. Their results give bounds on the depth of a ReLU net needed to compute exactly a piecewise linear function of d variables. However, except when $d=1$, they do not obtain an estimate on the number of neurons in such a network and hence cannot bound the width of the hidden layers.
4. Our results are related to Theorems II.1 and II.4 of Rolnick-Tegmark [16], which are themselves extensions of Lin-Rolnick-Tegmark [4]. Their results give lower bounds on the total size (number of neurons) of a neural net (with non-ReLU activations) that approximates sparse multivariable polynomials. Their bounds do not imply a control on the width of such networks that depends only on the number of variables, however.

5. This work was inspired in part by questions raised in the work of Telgarsky [8–10]. In particular, in Theorems 1.1 and 1.2 of [8], Telgarsky constructed interesting examples of sawtooth functions that can be computed efficiently by deep width 2 ReLU nets that cannot be well approximated by shallower networks with a similar number of parameters.
6. Theorems 1 and 2 are quantitative statements about the expressive power of depth without the aid of width. This topic, usually without considering bounds on the width, has been taken up by many authors. We refer the reader to [6,7] for several interesting quantitative measures of the complexity of functions computed by deep neural nets.
7. Finally, we refer the reader to the interesting work of Yarofsky [11], which provides bounds on the total number of parameters in a ReLU net needed to approximate a given class of functions (mainly balls in various Sobolev spaces).

4. Proof of Theorem 2

Proof of Theorem 2. We first treat the case:

$$f = \sup_{1 \leq \alpha \leq N} g_\alpha, \qquad g_\alpha : [0,1]^d \to \mathbb{R} \quad \text{affine}$$

when f is convex. We seek to show that f can be exactly represented by a ReLU net with input dimension d, hidden layer width $d+1$, and depth N. Our proof relies on the following observation.

Lemma 1. *Fix $d \geq 1$, and let $T : \mathbb{R}_+^d \to \mathbb{R}$ be an arbitrary function and $L : \mathbb{R}^d \to \mathbb{R}$ be affine. Define an invertible affine transformation $A : \mathbb{R}^{d+1} \to \mathbb{R}^{d+1}$ by:*

$$A(x,y) = (x, L(x) + y).$$

Then, the image of the graph of T under:

$$A \circ \mathrm{ReLU} \circ A^{-1}$$

is the graph of $x \mapsto \max\{T(x), L(x)\}$, viewed as a function on $\mathbb{R}_+^d$.

Proof. We have $A^{-1}(x,y) = (x, -L(x) + y)$. Hence, for each $x \in \mathbb{R}_+^d$, we have:

$$A \circ \mathrm{ReLU} \circ A^{-1}(x, T(x)) = \left(x, (T(x) - L(x)) \mathbf{1}_{\{T(x) - L(x) > 0\}} + L(x)\right)$$
$$= (x, \max\{T(x), L(x)\}).$$

□

We now construct a neural net that computes f. We note that the construction is potentially applicable to the study of avoiding sets (see the work of Shang [17]). Define invertible affine functions $A_\alpha : \mathbb{R}^{d+1} \to \mathbb{R}^{d+1}$ by:

$$A_\alpha(x, x_{d+1}) := (x, g_\alpha(x) + x_{d+1}), \qquad x = (x_1, \ldots, x_d),$$

and set:

$$H_\alpha := A_\alpha \circ \mathrm{ReLU} \circ A_\alpha^{-1}.$$

Further, define:

$$H_{\mathrm{out}} := \mathrm{ReLU} \circ \langle \vec{e}_{d+1}, \cdot \rangle \tag{13}$$

where $\vec{e}_{d+1}$ is the $(d+1)$th standard basis vector so that $\langle \vec{e}_{d+1}, \cdot \rangle$ is the linear map from $\mathbb{R}^{d+1}$ to $\mathbb{R}$ that maps $(x_1, \ldots, x_{d+1})$ to x_{d+1}. Finally, set:

$$H_{\text{in}} := \text{ReLU} \circ (\text{id}, 0),$$

where $(\text{id}, 0)(x) = (x, 0)$ maps $[0,1]^d$ to the graph of the zero function. Note that the ReLU in this initial layer is linear. With this notation, repeatedly using Lemma 1, we find that:

$$H_{\text{out}} \circ H_N \circ \cdots \circ H_1 \circ H_{\text{in}}$$

therefore has input dimension d, hidden layer width $d+1$, depth N, and computes f exactly.

Next, consider the general case when f is given by:

$$f = g - h, \qquad g = \sup_{1 \leq \alpha \leq N} g_\alpha, \qquad h = \sup_{1 \leq \beta \leq M} h_\beta$$

as in (11). For this situation, we use a different way of computing the maximum using ReLU nets.

Lemma 2. *There exists a ReLU net $\mathcal{M}$ with input dimension 2, hidden layer width 2, output dimension 1, and depth 2 such that:*

$$\mathcal{M}(x, y) = \max\{x, y\}, \qquad x \in \mathbb{R}, y \in \mathbb{R}_+.$$

Proof. Set $A_1(x, y) := (x - y, y)$, $A_2(z, w) = z + w$, and define:

$$\mathcal{M} = \text{ReLU} \circ A_2 \circ \text{ReLU} \circ A_1.$$

We have for each $y \geq 0, x \in \mathbb{R}$:

$$f_{\mathcal{M}}(x, y) = \text{ReLU}((x - y)\mathbf{1}_{\{x - y > 0\}} + y) = \max\{x, y\},$$

as desired. □

We now describe how to construct a ReLU net $\mathcal{N}$ with input dimension d, hidden layer width $d + 3$, output dimension 1, and depth $2(M + N)$ that exactly computes f. We use width d to copy the input x, width 2 to compute successive maximums of the positive affine functions g_α, h_β using the net $\mathcal{M}$ from Lemma 2 above, and width 1 as memory in which we store $g = \sup_\alpha g_\alpha$ while computing $h = \sup_\beta h_\beta$. The final layer computes the difference $f = g - h$. □

5. Proof of Theorem 1

Proof of Theorem 1. We begin by showing (8) and (9). Suppose $f : [0,1]^d \to \mathbb{R}_+$ is convex, and fix $\varepsilon > 0$. A simple discretization argument shows that there exists a piecewise affine convex function $g : [0,1]^d \to \mathbb{R}_+$ such that $\|f - g\|_{C^0} \leq \varepsilon$. By Theorem 2, g can be exactly represented by a ReLU net with hidden layer width $d + 1$. This proves (8). In the case that f is Lipschitz, we use the following, a special case of Lemma 4.1 in [15].

Proposition 1. *Suppose $f : [0,1]^d \to \mathbb{R}$ is convex and Lipschitz with Lipschitz constant L. Then, for every $k \geq 1$, there exist k affine maps $A_j : [0,1]^d \to \mathbb{R}$ such that:*

$$\left\| f - \sup_{1 \leq j \leq k} A_j \right\|_{C^0} \leq 72L\, d^{3/2} k^{-2/d}.$$

Combining this result with Theorem 2 proves (9). We turn to checking (5) and (10). We need the following observations, which seems to be well known, but not written down in the literature.

Lemma 3. *Let $\mathcal{N}$ be a ReLU net with input dimension d, a single hidden layer of width n, and output dimension 1. There exists another ReLU net $\widetilde{\mathcal{N}}$ that computes the same function as $\mathcal{N}$, but has input dimension d and $n+2$ hidden layers with width $d+2$.*

Proof. Denote by $\{A_j\}_{j=1}^n$ the affine functions computed by each neuron in the hidden layer of $\mathcal{N}$ so that:
$$f_{\mathcal{N}}(x) = \mathrm{ReLU}\left(b + \sum_{j=1}^n c_j \mathrm{ReLU}(A_j(x))\right).$$

Let $T > 0$ be sufficiently large so that:
$$T + \sum_{j=1}^k c_j \mathrm{ReLU}(A_j(x)) > 0, \qquad \forall 1 \leq k \leq n, \ x \in [0,1]^d.$$

The affine transformations $\widetilde{A}_j$ computed by the jth hidden layer of $\widetilde{\mathcal{N}}$ are then:
$$\widetilde{A}_1(x) := (x, A_j(x), T) \quad \text{and} \quad \widetilde{A}_{n+2}(x,y,z) = z - T + b, \qquad x \in \mathbb{R}^d, y, z \in \mathbb{R}$$

and:
$$\widetilde{A}_j(x,y,z) = (x, A_j(x), z + c_{j-1}y), \qquad j = 2, \ldots, n+1.$$

We are essentially using width d to copy in the input variable, width 1 to compute each A_j, and width 1 to store the output. □

Recall that positive continuous functions can be arbitrarily well approximated by smooth functions and hence by ReLU nets with a single hidden layer (see, e.g., Theorem 3.1 [5]). The relation (5) therefore follows from Lemma 3. Similarly, by Theorem 3.1 in [5], if f is smooth, then there exists $K = K(d) > 0$ and a constant C_f depending only on the maximum value of the first K derivatives of f such that:
$$\inf_{\mathcal{N}} \|f - f_{\mathcal{N}}\| \leq C_f n^{-1/d},$$

where the infimum is over ReLU nets $\mathcal{N}$ with a single hidden layer of width n. Combining this with Lemma 3 proves (10).

It remains to prove (6) and (7). To do this, fix a positive continuous function $f : [0,1]^d \to \mathbb{R}_+$ with modulus of continuity ω_f. Recall that the volume of the unit d-simplex is $1/d!$, and fix $\varepsilon > 0$. Consider the partition:
$$[0,1]^d = \bigcup_{j=1}^{d!/\omega_f(\varepsilon)^d} \mathcal{P}_j$$

of $[0,1]^d$ into $d!/\omega_f(\varepsilon)^d$ copies of $\omega_f(\varepsilon)$ times the standard d-simplex. Here, each $\mathcal{P}_j$ denotes a single scaled copy of the unit simplex. To create this partition, we first sub-divide $[0,1]^d$ into at most $\omega_f(\varepsilon)^{-d}$ cubes of side length at most $\omega_f(\varepsilon)$. Then, we subdivide each such smaller cube into $d!$ copies of the standard simplex (which has volume $1/d!$) rescaled to have side length $\omega_f(\varepsilon)$. Define f_ε to be a piecewise linear approximation to f obtained by setting f_ε equal to f on the vertices of the $\mathcal{P}_j$'s and taking f_ε to be affine on their interiors. Since the diameter of each $\mathcal{P}_j$ is $\omega_f(\varepsilon)$, we have:
$$\|f - f_\varepsilon\|_{C^0} \leq \varepsilon.$$

Next, since f_ε is a piecewise affine function, by Theorem 2.1 in [2] (see Theorem 2), we may write:

$$f_\varepsilon = g_\varepsilon - h_\varepsilon,$$

where $g_\varepsilon, h_\varepsilon$ are convex, positive, and piecewise affine. Applying Theorem 2 completes the proof of (6) and (7). □

6. Conclusions

We considered in this article the expressive power of ReLU networks with bounded hidden layer widths. In particular, we showed that ReLU networks of width $d+3$ and arbitrary depth are capable of arbitrarily good approximations of any scalar continuous function of d variables. We showed further that this bound could be reduced to $d+1$ in the case of convex functions and gave quantitative rates of approximation in all cases. Our results show that deep ReLU networks, even at a moderate width, are universal function approximators. Our work leaves open the question of whether such function representations can be learned by (stochastic) gradient descent from a random initialization. We will take up this topic in future work.

Funding: This research was funded by NSF Grants DMS-1855684 and CCF-1934904.

Acknowledgments: It is a pleasure to thank Elchanan Mossel and Leonid Hanin for many helpful discussions. This paper originated while I attended EM's class on deep learning [18]. In particular, I would like to thank him for suggesting proving quantitative bounds in Theorem 2 and for suggesting that a lower bound can be obtained by taking piece-wise linear functions with many different directions. He also pointed out that the width estimates for the continuous function in Theorem 1 were sub-optimal in a previous draft. I would also like to thank Leonid Hanin for detailed comments on several previous drafts and for useful references to the results in approximation theory. I am also grateful to Brandon Rule and Matus Telgarsky for comments on an earlier version of this article. I am also grateful to BR for the original suggestion to investigate the expressivity of neural nets of width two. I also would like to thank Max Kleiman-Weiner for useful comments and discussion. Finally, I thank Zhou Lu for pointing out a serious error what used to be Theorem 3 in a previous version of this article. I have removed that result.

Conflicts of Interest: The authors declare no conflict of interest.

References

1. Bengio, Y.; Hinton, G.; LeCun, Y. Deep learning. *Nature* **2015**, *521*, 436–444.
2. Arora, R.; Basu, A.; Mianjy, P.; Mukherjee, A. Understanding deep neural networks with Rectified Linear Units. In Proceedings of the International Conference on Representation Learning, Vancouver, BC, Canada, 30 April 30–3 May 2018.
3. Liao, Q.; Mhaskar, H.; Poggio, T. Learning functions: When is deep better than shallow. *arXiv* **2016**, arXiv:1603.00988v4.
4. Lin, H.; Rolnick, D.; Tegmark, M. Why does deep and cheap learning work so well? *arXiv* **2016**, arXiv:1608.08225v3.
5. Mhaskar, H.; Poggio, T. Deep vs. shallow networks: An approximation theory perspective *Anal. Appl.* **2016**, *14*, 829–848. [CrossRef]
6. Poole, B.; Lahiri, S.; Raghu, M.; Sohl-Dickstein, J.; Ganguli, S. Exponential expressivity in deep neural networks through transient chaos. *Adv. Neural Inf. Process. Syst.* **2016**, *29*, 3360–3368.
7. Raghu, M.; Poole, B.; Kleinberg, J.; Ganguli, S.; Dickstein, J. On the expressive power of deep neural nets. In Proceedings of the 34th International Conference on Machine Learning, Sydney, Australia, 6–11 August 2017; Volume 70, pp. 2847–2854.
8. Telgrasky, M. Representation benefits of deep feedforward networks. *arXiv* **2015**, arXiv:1509.08101.
9. Telgrasky, M. Benefits of depth in neural nets. In Proceedings of the JMLR: Workshop and Conference Proceedings, New York, NY, USA, 19 June 2016; Volume 49, pp. 1–23.
10. Telgrasky, M. Neural networks and rational functions. In Proceedings of the 34th International Conference on Machine Learning, Sydney, Australia, 6–11 August 2017; Volume 70, pp. 3387–3393.

11. Yarotsky, D. Error bounds for approximations with deep ReLU network. *Neural Netw.* **2017**, *94*, 103–114. [CrossRef] [PubMed]
12. Cybenko, G. Approximation by superpositions of a sigmoidal function. *Math. Control. Signals Syst. (MCSS)* **1989**, *2*, 303–314. [CrossRef]
13. Hornik, K.; Stinchcombe, M.; White, H. Multilayer feedforward networks are universal approximators. *J. Neural Netw.* **1989**, *2*, 359–366 [CrossRef]
14. Hanin, B.; Sellke, M. Approximating Continuous Functions by ReLU Nets of Minimal Width. *arXiv* **2017**, arXiv:1710.11278.
15. Balázs, G.; György, A.; Szepesvári, C. Near-optimal max-affine estimators for convex regression. In Proceedings of the Eighteenth International Conference on Artificial Intelligence and Statistics, San Diego, CA, USA, 9–12 May 2015; Volume 38, pp. 56–64.
16. Rolnick, D.; Tegmark, M. The power of deeper networks for expressing natural functions. In Proceedings of International Conference on Representation Learning, Vancouver, BC, Canada, 30 April–3 May 2018.
17. Shang, Y. A combinatorial necessary and sufficient condition for cluster consensus. *Neurocomputing* **2016**, *216*, 611–616. [CrossRef]
18. Mossel, E. Mathematical Aspects of Deep Learning. Available online: http://elmos.scripts.mit.edu/mathofdeeplearning/mathematical-aspects-of-deep-learning-intro/ (accessed on 10 September 2019)

© 2019 by the author. Licensee MDPI, Basel, Switzerland. This article is an open access article distributed under the terms and conditions of the Creative Commons Attribution (CC BY) license (http://creativecommons.org/licenses/by/4.0/).

Article

Prediction of Discretization of GMsFEM Using Deep Learning

Min Wang [1], Siu Wun Cheung [1], Eric T. Chung [2], Yalchin Efendiev [1,3,*], Wing Tat Leung [4] and Yating Wang [1,5]

1. Department of Mathematics, Texas A&M University, College Station, TX 77843, USA; wangmin@math.tamu.edu (M.W.); tonycsw2905@math.tamu.edu (S.W.C.); wytgloria@math.tamu.edu (Y.W.)
2. Department of Mathematics, The Chinese University of Hong Kong, Hong Kong, China; tschung@math.cuhk.edu.hk
3. Multiscale Model Reduction Laboratory, North-Eastern Federal University, 677980 Yakutsk, Russia
4. Institute for Computational Engineering and Sciences, The University of Texas at Austin, Austin, TX 78712, USA; sidnet123@gmail.com
5. Department of Mathematics, Purdue University, West Lafayette, IN 47907, USA
* Correspondence: efendiev@math.tamu.edu

Received: 25 March 2019; Accepted: 30 April 2019; Published: 8 May 2019

Abstract: In this paper, we propose a deep-learning-based approach to a class of multiscale problems. The generalized multiscale finite element method (GMsFEM) has been proven successful as a model reduction technique of flow problems in heterogeneous and high-contrast porous media. The key ingredients of GMsFEM include mutlsicale basis functions and coarse-scale parameters, which are obtained from solving local problems in each coarse neighborhood. Given a fixed medium, these quantities are precomputed by solving local problems in an offline stage, and result in a reduced-order model. However, these quantities have to be re-computed in case of varying media (various permeability fields). The objective of our work is to use deep learning techniques to mimic the nonlinear relation between the permeability field and the GMsFEM discretizations, and use neural networks to perform fast computation of GMsFEM ingredients repeatedly for a class of media. We provide numerical experiments to investigate the predictive power of neural networks and the usefulness of the resultant multiscale model in solving channelized porous media flow problems.

Keywords: generalized multiscale finite element method; multiscale model reduction; deep learning

1. Introduction

Multiscale features widely exist in many engineering problems. For instance, in porous media flow, the media properties typically vary over many scales and contain high contrast. Multiscale finite element methods (MsFEM) [1–3] and generalized multiscale finite element methods (GMsFEM) [4,5] are designed for solving multiscale problems using local model reduction techniques. In these methods, the computational domain is partitioned into a coarse grid $\mathcal{T}^H$, which does not necessarily resolve all multiscale features. We further perform a refinement of $\mathcal{T}^H$ to obtain a fine grid $\mathcal{T}^h$, which essentially resolves all multiscale features. The idea of local model reduction in these methods is based on idenfications of local multiscale basis functions supported in coarse regions on the fine grid, and replacement of the macroscopic equations by a coarse-scale system using a limited number of local multiscale basis functions. As in many model reduction techniques, the computations of multiscale basis functions, which constitute a small dimensional subspace, can be performed in an offline stage. For a fixed medium, these multiscale basis functions are reusable for any force terms and boundary conditions. Therefore, these methods provide substantial computational savings in the online stage, in which a coarse-scale system is constructed and solved on the reduced-order space.

However, difficulties arise in situations with uncertainties in the media properties in some local regions, which are common for oil reservoirs or aquifers. One straightforward approach for quantifying the uncertainties is to sample realizations of media properties. In such cases, it is challenging to find an offline principal component subspace which is able to universally solve the multiscale problems with different media properties. The computation of multiscale basis functions has to be performed in an online procedure for each medium. Even though the multiscale basis functions are reusable for different force terms and boundary conditions, the computational effort can grow very huge for a large number of realizations of media properties. To this end, building a functional relationship between the media properties and the multiscale model in an offline stage can avoid repeating expensive computations and thus vastly reduce the computational complexity. Due to the diversity of complexity of the media properties, the functional relationship is highly nonlinear. Modelling such a nonlinear functional relationship typically involves high-order approximations. Therefore, it is natural to use machine learning techniques to devise such complex models. In [6,7], the authors make use of a Bayesian approach for learning multiscale models and incorporating essential observation data in the presence of uncertainties.

Deep neural networks is one class of machine learning algorithm that is based on an artificial neural network, which is composed of a relatively large number of layers of nonlinear processing units, called neurons, for feature extraction. The neurons are connected to other neurons in the successive layers. The information propagates from the input, through the intermediate hidden layers, and to the output layer. In the propagation process, the output in each layer is used as input in the consecutive layer. Each layer transforms its input data into a little more abstract feature representation. In between layers, a nonlinear activation function is used as the nonlinear transformation on the input, which increases the expressive power of neural networks. Recently, deep neural network (DNN) has been successfully used to interpret complicated data sets and applied to tasks with pattern recognition, such as image recognition, speech recognition and natural language processing [8–10]. Extensive researches have also been conducted on investigating the expression power of deep neural networks [11–15].

Results show that neural networks can represent and approximate a large class of functions. Recently, deep learning has been applied to model reductions and partial differential equations. In [16], the authors studied deep convolution networks for surrogate model construction. on dynamic flow problems in heterogeneous media. In [17], the authors studied the relationship between residual networks (ResNet) and characteristic equations of linear transport, and proposed an interpretation of deep neural networks by continuous flow models. In [18], the authors combined the idea of the Ritz method and deep learning techniques to solve elliptic problems and eigenvalue problems. In [19], a neural network has been designed to learn the physical quantities of interest as a function of random input coefficients. The concept of using deep learning to generate a reduced-order model for a dynamic flow has been applied to proper orthogonal decomposition (POD) global model reduction [20] and nonlocal multi-continuum upscaling (NLMC) [21].

In this work, we propose a deep-learning-based method for fast computation of the GMsFEM discretization. Our approach makes use of deep neural networks as a fast proxy to compute GMsFEM discretizations for flow problems in channelized porous media with uncertainties. More specifically, neural networks are used to express the functional relationship between the media properties and the multiscale model. Such networks are built up in an offline stage. Sufficient sample pairs are required to ensure the expressive power of the networks. With different realizations of media properties, one can use the built network and avoid computations of local problems and spectral problems.

The paper is organized as follows. We start with the underlying partial differential equation that describes the flow within a heterogeneous media and the main ingredients of GMsFEM in Section 2. Next, in Section 3, we present the idea of using deep learning as a proxy for prediction of GMsFEM discretizations. The networks will be precisely defined and the sampling will be explained in detail.

In Section 4, we present numerical experiments to show the effectiveness of our presented networks on several examples with different configurations. Finally, a conclusive discussion is provided in Section 5.

2. Preliminaries

In this paper, we are considering the flow problem in highly heterogeneous media

$$-\mathrm{div}(\kappa \nabla u) = f \quad \text{in} \quad \Omega, \qquad (1)$$
$$u = 0 \quad \text{or} \quad \frac{\partial u}{\partial n} = 0 \quad \text{on} \quad \partial\Omega,$$

where Ω is the computational domain, κ is the permeability coefficient in $L^\infty(\Omega)$, and f is a source function in $L^2(\Omega)$. We assume the coefficient κ is highly heterogeneous with high contrast. The classical finite element method for solving (1) numerically is given by: find $u_h \in V_h$ such that

$$a(u_h, v) = \int_\Omega \kappa \nabla u_h \cdot \nabla v \, dx = \int_\Omega f v \, dx = (f, v) \quad \text{for all } v \in V_h, \qquad (2)$$

where V_h is a standard conforming finite element space over a partition $\mathcal{T}_h$ of Ω with mesh size h.

However, with the highly heterogeneous property of coefficient κ, the mesh size h has to be taken extremely small to capture the underlying fine-scale features of κ. This ends up with a large computational cost. GMsFEM [4,5] serves as a model reduction technique to reduce the number of degrees of freedom and attain both efficiency and accuracy to a considerable extent. GMsFEM has been successfully extended to other formulations and applied to other problems. Here we provide a brief introduction of the main ingredients of GMsFEM. For a more detailed discussion of GMsFEM and related concepts, the reader is referred to [22–26].

In GMsFEM, we define a coarse mesh $\mathcal{T}^H$ over the domain Ω and refine to obtain a fine mesh $\mathcal{T}^h$ with mesh size $h \ll H$, which is fine enough to restore the multiscale properties of the problem. Multiscale basis functions are defined on coarse grid blocks using linear combinations of finite element basis functions on $\mathcal{T}^h$, and designed to resolve the local multiscale behaviors of the exact solution. The multiscale finite element space V_{ms}, which is a principal component subspace of the conforming finite space V_h with $\dim(V_{\mathrm{ms}}) \ll \dim(V_h)$, is constructed by the linear span of multiscale basis functions. The multiscale solution $u_{\mathrm{ms}} \in V_{\mathrm{ms}}$ is then defined by

$$a(u_{\mathrm{ms}}, v) = (f, v) \quad \text{for all } v \in V_{\mathrm{ms}}. \qquad (3)$$

We consider the identification of dominant modes for solving (1) by multiscale basis functions, including spectral basis functions and simplified basis functions, in GMsFEM. Here, we present the details of the construction of multiscale basis functions in GMsFEM. Let $N_x = \{x_i \mid 1 \leq i \leq N_v\}$ be the set of nodes of the coarse mesh $\mathcal{T}^H$. For each coarse grid node $x_i \in N_x$, the coarse neighborhood ω_i is defined by

$$\omega_i = \bigcup \{K_j \in \mathcal{T}^H; \ x_i \in \overline{K}_j\}, \qquad (4)$$

that is, the union of the coarse elements $K_j \in \mathcal{T}^H$ containing the coarse grid node x_i. An example of the coarse and fine mesh, coarse blocks and a coarse neighborhood is shown in Figure 1. For each coarse neighbourhood ω_i, we construct multiscale basis functions $\{\phi_j^{\omega_i}\}_{j=1}^{L_i}$ supported on ω_i.

For the construction of spectral basis functions, we first construct a snapshot space $V_{\mathrm{snap}}^{(i)}$ spanned by local snapshot basis functions $\psi_{\mathrm{snap}}^{i,k}$ for each local coarse neighborhood ω_i. The snapshot basis function $\psi_{\mathrm{snap}}^{i,k}$ is the solution of a local problem

$$-\mathrm{div}(\kappa \nabla \psi_{\mathrm{snap}}^{i,k}) = 0, \quad \text{in} \quad \omega_i \qquad (5)$$
$$\psi_{\mathrm{snap}}^{i,k} = \delta_{i,k}, \quad \text{on} \quad \partial \omega_i.$$

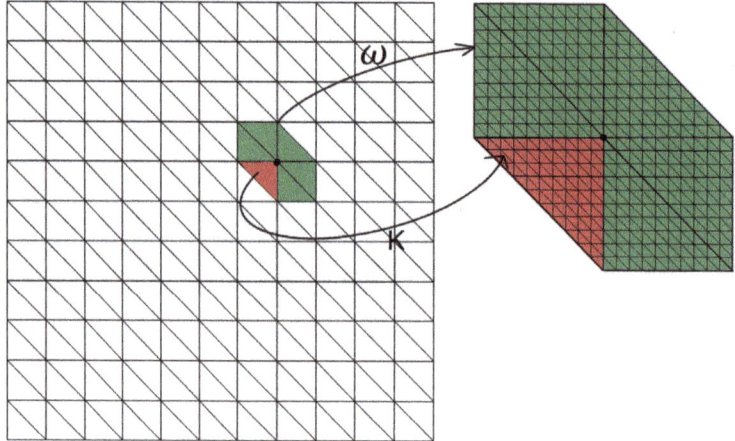

Figure 1. An illustration of coarse mesh (**left**), a coarse neighborhood and coarse blocks (**right**).

The fine grid function $\delta_{i,k}$ is a function defined for all $x_s \in \partial \omega_i$, where $\{x_s\}$ denote the fine degrees of freedom on the boundary of local coarse region ω_i. Specifically,

$$\delta_{i,k}(x_s) = \begin{cases} 1 & \text{if } s = k \\ 0 & \text{if } s \neq k. \end{cases} \tag{6}$$

The linear span of these harmonic extensions forms the local snapshot space $V_{\text{snap}}^{(i)} = \text{span}_k\{\psi_{\text{snap}}^{i,k}\}$. One can also use randomized boundary conditions to reduce the computational cost associated with snapshot calculations [25]. Next, a spectral problem is designed based on our analysis and used to reduce the dimension of local multiscale space. More precisely, we seek for eigenvalues λ_m^i and corresponding eigenfunctions $\phi_m^{\omega_i} \in V_{\text{snap}}^{(i)}$ satisfying

$$a_i(\phi_m^{\omega_i}, v) = \lambda_m^i s_i(\phi_m^{\omega_i}, v), \quad \forall v \in V_{\text{snap}}^{(i)}, \tag{7}$$

where the bilinear forms in the spectral problem are defined as

$$\begin{aligned} a_i(u,v) &= \int_{\omega_i} \kappa \nabla u \cdot \nabla v, \\ s_i(u,v) &= \int_{\omega_i} \tilde{\kappa} u v, \end{aligned} \tag{8}$$

where $\tilde{\kappa} = \sum_j \kappa |\nabla \chi_j|^2$, and χ_j denotes the multiscale partition of the unity function. We arrange the eigenvalues λ_m^i of the spectral problem (7) in ascending order, and select the first l_i eigenfunctions $\{\phi_m^{\omega_i}\}_{m=1}^{l_i}$ corresponding to the small eigenvalues as the multiscale basis functions.

An alternative way to construct the multiscale basis function is by using the idea of simplified basis functions. This approach assumes the number of channels and position of the channelized permeability field are known. Therefore we can obtain multiscale basis functions $\{\phi_m^{\omega_i}\}_{m=1}^{l_i}$ using these information and without solving the spectral problem [27].

Once the multiscale basis functions are constructed, the span of the multiscale basis functions will form the offline space

$$\begin{aligned} V_{ms}^{(i)} &= \text{span}\{\phi_m^{\omega_i}\}_{m=1}^{l_i}, \\ V_{ms} &= \oplus_i V_{ms}^{(i)}. \end{aligned} \tag{9}$$

We will then seek a multlscale solution $u_{ms} \in V_{ms}$ satisfying

$$a(u_{ms}, v) = (f, v) \quad \text{for all } v \in V_{ms}, \tag{10}$$

which is a Galerkin projection of the (1) onto V_{ms}, and can be written as a system of linear equations

$$A_c u_c = b_c, \tag{11}$$

where A_c and b_c are the coarse-scale stiffness matrix and load vector. If we collect all the multiscale basis functions and arrange the fine-scale coordinate representation in columns, we obtain the downscaling operator R. Then the fine-scale representation of the multiscale solution is given by

$$u_{ms} = R u_c. \tag{12}$$

3. Deep Learning for GMsFEM

In applications, there are uncertainties within some local regions of the permeability field κ in the flow problem. Thousands of forward simulations are needed to quantify the uncertainties of the flow solution. GMsFEM provides us with a fast solver to compute the solutions accurately and efficiently. Considering that there is a large amount of simulation data, we are interested in developing a method utilizing the existing offline data and reducing direct computational effort later. In this work, we aim at using DNN to model the relationship between heterogeneous permeability coefficient κ and the key ingredients of GMsFEM solver, i.e., coarse scale stiffness matrices and multiscale basis functions. When the relation is built up, we can feed the network any realization of the permeability field and obtain the corresponding GMsFEM ingredients, and further restore fine-grid GMsFEM solution of (1). The general idea of utilizing deep learning in the GMsFEM framework is illustrated in Figure 2.

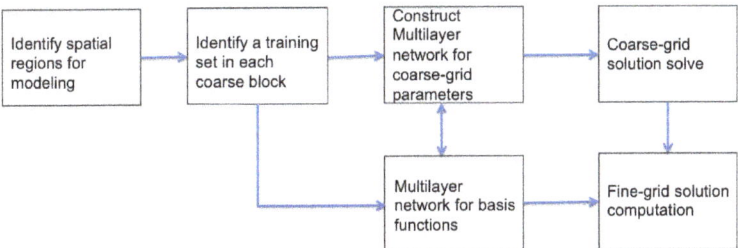

Figure 2. A flow chart in illustrating the idea of using deep learning in the generalized multiscale finite element method (GMsFEM) framework.

Suppose that there are uncertainties for the heterogeneous coefficient in a local coarse block K_0, which we call the target block, and the permeability outside the target block remains the same. For example, for a channelized permeability field, the position, location and the permeability values of the channels in the target block can vary. The target block K_0 is inside three coarse neighborhoods, denoted by $\omega_1, \omega_2, \omega_3$. The union of the 3 neighborhoods, i.e.,

$$\omega^+(K_0) = \omega_1 \cup \omega_2 \cup \omega_3, \tag{13}$$

are constituted of by the target block K_0 and 12 other coarse blocks, denoted by $\{K_l\}_{l=1}^{12}$ A target block and its surrounding neighborhoods are depicted in Figure 3.

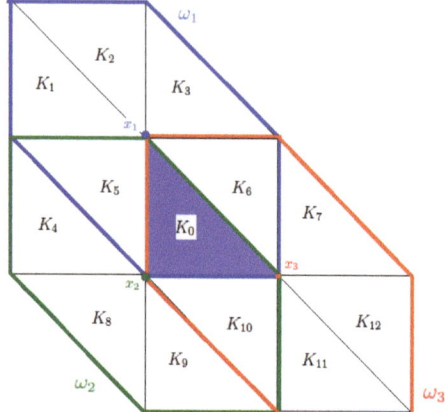

Figure 3. An illustration of a target coarse block K_0 and related neighborhoods.

For a fixed permeability field κ, one can compute the multiscale basis functions $\phi_m^{\omega_i}(\kappa)$ defined by (7), for $i = 1, 2, 3$, and the local coarse-scale stiffness matrices $A_c^{K_l}(\kappa)$, defined by

$$[A_c^{K_l}(\kappa)]_{m,n}^{i,j} = \int_{K_l} \kappa \nabla \phi_m^{\omega_i}(\kappa) \cdot \nabla \phi_n^{\omega_j}(\kappa), \tag{14}$$

for $l = 0, 1, \ldots, 12$. We are interested in constructing the maps $g_B^{m,i}$ and g_M^l, where

- $g_B^{m,i}$ maps the permeability coefficient κ to a local multiscale basis function $\phi_m^{\omega_i}$, where i denotes the index of the coarse block, and m denotes the index of the basis in coarse block ω_i

$$g_B^{m,i} : \kappa \mapsto \phi_m^{\omega_i}(\kappa), \tag{15}$$

- g_M^l maps the permeability coefficient κ to the coarse grid parameters $A_c^{K_l}$ ($l = 0, \cdots, 12$)

$$g_M^l : \kappa \mapsto A_c^{K_l}(\kappa). \tag{16}$$

In this work, our goal is to make use of deep learning to build fast approximations of these quantities associated with the uncertainties in the permeability field κ, which can provide fast and accurate solutions to the heterogeneous flow problem (1).

For each realization κ, one can compute the images of κ under the local multiscale basis maps $g_B^{m,i}$ and the local coarse-scale matrix maps g_M^l. These forward calculations serve as training samples for building a deep neural network for approximation of the corresponding maps, i.e.,

$$\begin{aligned} \mathcal{N}_B^{m,i}(\kappa) &\approx g_B^{m,i}(\kappa), \\ \mathcal{N}_M^l(\kappa) &\approx g_M^l(\kappa). \end{aligned} \tag{17}$$

In our networks, the permeability field κ is the input, while the multiscale basis functions $\phi_m^{\omega_i}$ and the coarse-scale matrix $A_c^{K_l}$ are the outputs. Once the neural networks are built, we can use the networks to compute the multiscale basis functions and coarse-scale parameters in the associated region for any permeability field κ. Using these local information from the neural networks together with the global information which can be pre-computed, we can form the downscale operator R with the multiscale basis functions, form and solve the linear system (11), and obtain the multiscale solution by (12).

3.1. Network Architecture

In general, an L-layer neural network $\mathcal{N}$ can be written in the form

$$\mathcal{N}(x;\theta) = \sigma(W_L\sigma(\cdots\sigma(W_2\sigma(W_1x+b_1)+b_2)\cdots)+b_L), \tag{18}$$

where $\theta := (W_1, W_2, \cdots, W_L, b_1, b_2, \cdots, b_L)$, W's are the weight matrices and b's are the bias vectors, σ is the activation function, x is the input. Such a network is used to approximate the output y. Our goal is then to find θ^* by solving an optimization problem

$$\theta^* = \underset{\theta}{\mathrm{argmin}}\ \mathcal{L}(\theta), \tag{19}$$

where $\mathcal{L}(\theta)$ is called loss function, which measures the mismatch between the image of the input x under the the neural network $\mathcal{N}(x,y;\theta)$ and the desired output y in a set of training samples (x_j, y_j). In this paper, we use the mean-squared error metric to be our loss function

$$\mathcal{L}(\theta) = \frac{1}{N}\sum_{j=1}^{N}\|y_j - \mathcal{N}(x_j;\theta)\|_2^2, \tag{20}$$

where N is the number of the training samples. An illustration of a deep neural network is shown in Figure 4.

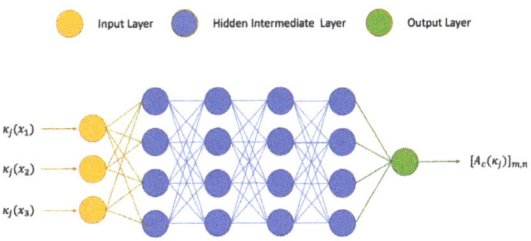

Figure 4. An illustration of a deep neural network.

Suppose we have a set of different realizations of the permeability $\{\kappa_1, \kappa_2, \cdots, \kappa_N\}$ in the target block. In our network, the input $x_j = \kappa_j \in \mathbb{R}^d$ is a vector containing the permeability image pixels in the target block. The output y_j is an entry of the local stiffness matrix $A_c^{K_l}$, or the coordinate representation of a multiscale basis function $\phi_m^{\omega_i}$. We will make use of these sample pairs (x_j, y_j) to train a deep neural network $\mathcal{N}_B^{m,i}(x;\theta_B^*)$ and $\mathcal{N}_M^l(x;\theta_M^*)$ by minimizing the loss function with respect to the network parameter θ, such that the trained neural networks can approximate the functions $g_B^{m,i}$ and g_M^l, respectively. Once the neural is constructed, for some given new permeability coefficient κ_{N+1}, we use our trained networks to compute a fast prediction of the outputs, i.e., local multiscale basis functions $\phi_m^{\omega_i,\mathrm{pred}}$ by

$$\phi_m^{\omega_i,\mathrm{pred}}(\kappa_{N+1}) = \mathcal{N}_B^{m,i}(\kappa_{N+1};\theta_B^*) \approx g_B^{m,i}(\kappa_{N+1}) = \phi_m^{\omega_i}(\kappa_{N+1}), \tag{21}$$

and local coarse-scale stiffness matrix $A_c^{K_l,\mathrm{pred}}$ by

$$A_c^{K_l,\mathrm{pred}}(\kappa_{N+1}) = \mathcal{N}_M^l(\kappa_{N+1};\theta_M^*) \approx g_M^l(\kappa_{N+1}) = A_c^{K_l}(\kappa_{N+1}). \tag{22}$$

3.2. Network-Based Multiscale Solver

Once the neural networks are built, we can assemble the predicted multiscale basis functions to obtain a prediction R^{pred} for the downscaling operator, and assemble the predicted local coarse-scale stiffness matrix $A_c^{K_l,\text{pred}}$ in the global matrix A_c^{pred}. Following (11) and (12), we solve the predicted coarse-scale coefficient vector u_c^{pred} from the following linear system

$$A_c^{\text{pred}} u_c^{\text{pred}} = b_c, \qquad (23)$$

and obtain the predicted multiscale solution u_{ms}^{pred} by

$$u_{ms}^{\text{pred}} = R^{\text{pred}} u_c^{\text{pred}}. \qquad (24)$$

4. Numerical Results

In this section, we present some numerical results for predicting the GMsFEM ingredients and solutions using our proposed method. We considered permeability fields κ with high-contrast channels inside the domain $\Omega = (0,1)^2$, which consist of uncertainties in a target cell K_0. More precisely, we considered a number of random realizations of permeability fields $\kappa_1, \kappa_2, \kappa_3, \cdots, \kappa_{N+M}$. Each permeability field contained two high-conductivity channels, and the fields differ in the target cell K_0 by:

- in experiment 1, the channel configurations were all distinct, and the permeability coefficients inside the channels were fixed in each sample (see Figure 5 for illustrations), and
- in experiment 2, the channel configurations were randomly chosen among five configurations, and the permeability coefficients inside the channels followed a random distribution (see Figure 6 for illustrations).

In these numerical experiments, we assumed there were uncertainties in only the target block K_0. The permeability field in $\Omega \setminus K_0$ was fixed across all the samples.

We followed the procedures in Section 3 and generated sample pairs using GMsFEM. Local spectral problems were solved to obtain the multiscale basis functions $\phi_m^{\omega_i}$. In the neural network, the permeability field $x = \kappa$ was considered to be the input, while the local multiscale basis functions $y = \phi_m^{\omega_i}$ and local coarse-scale matrices $y = A_c^{K_l}$ were regarded as the output. These sample pairs were divided into the training set and the learning set in a random manner. A large number N of realizations, namely $\kappa_1, \kappa_2, \ldots, \kappa_N$, were used to generate sample pairs in the training set, while the remaining M realizations, namely $\kappa_{N+1}, \kappa_{N+2}, \ldots, \kappa_{N+M}$ are used in testing the predictive power of the trained network. We remark that, for each basis function and each local matrix, we solved an optimization problem in minimizing the loss function defined by the sample pairs in the training set, and build a separate deep neural network. We summarize the network architectures for training local coarse scale stiffness matrix and multiscale basis functions as below:

- For the multiscale basis function $\phi_m^{\omega_i}$, we built a network $\mathcal{N}_B^{m,i}$ using
    - Input: vectorized permeability pixels values κ,
    - Output: coefficient vector of multiscale basis $\phi_m^{\omega_i}(\kappa)$ on coarse neighborhood ω_i,
    - Loss function: mean squared error $\frac{1}{N} \sum_{j=1}^{N} ||\phi_m^{\omega_i}(\kappa_j) - \mathcal{N}_B^{m,i}(\kappa_j; \theta_B)||_2^2$,
    - Activation function: leaky ReLu function,
    - DNN structure: 10–20 hidden layers, each layer have 250–350 neurons,
    - Training optimizer: Adamax.
- For the local coarse scale stiffness matrix $A_c^{K_l}$, we build a network $\mathcal{N}_M^l$ using

- Input: vectorized permeability pixels values κ,
- Output: vectorized coarse scale stiffness matrix $A_c^{K_l}(\kappa)$ on the coarse block K_l,
- Loss function: mean squared error $\frac{1}{N}\sum_{j=1}^{N}||A_c^{K_l}(\kappa_j) - \mathcal{N}_M^l(\kappa_j; \theta_M)||_2^2$,
- Activation function: ReLu function (rectifier),
- DNN structure: 10–16 hidden layers, each layer have 100–500 neurons,
- Training optimizer: Proximal Adagrad.

For simplicity, the activation functions ReLU function [28] and Leaky ReLU function were used as they have the simplest derivatives among all nonlinear functions. The ReLU function proved to be useful in training deep neural network architectures. The Leaky ReLU function can resolve the vanishing gradient problem which can accelerate the training in some occasions. The optimizers Adamax and Proximal Adagrad are stochastic gradient descent (SGD)-based methods commonly used in neural network training [29]. In both experiments, we trained our network using Python API Tensorflow and Keras [30].

Once a neural network was built on training, it can be used to predict the output given a new input. The accuracy of the predictions is essential in making the network useful. In our experiments, we used M sample pairs, which were not used in training the network, to examine the predictive power of our network. On these sample pairs, referred to as the testing set, we compared the prediction and the exact output and computed the mismatch in some suitable metric. Here, we summarize the metric used in our numerical experiment. For the multiscale basis functions, we compute the relative error in L^2 and H^1 norm, i.e.,

$$e_{L^2}(\kappa_{N+j}) = \left(\frac{\int_\Omega |\phi_m^{\omega_i}(\kappa_{N+j}) - \phi_m^{\omega_i,\text{pred}}(\kappa_{N+j})|^2}{\int_\Omega |\phi_m^{\omega_i}(\kappa_{N+j})|^2}\right)^{\frac{1}{2}},$$

$$e_{H^1}(\kappa_{N+j}) = \left(\frac{\int_\Omega |\nabla\phi_m^{\omega_i}(\kappa_{N+j}) - \nabla\phi_m^{\omega_i,\text{pred}}(\kappa_{N+j})|^2}{\int_\Omega |\nabla\phi_m^{\omega_i}(\kappa_{N+j})|^2}\right)^{\frac{1}{2}}.$$
(25)

For the local stiffness matrices, we computed the relative error in entrywise ℓ^2, entrywise ℓ^∞ and Frobenius norm, i.e.,

$$e_{\ell^2}(\kappa_{N+j}) = \frac{||A_c^{K_l}(\kappa_{N+j}) - A_c^{K_l,\text{pred}}(\kappa_{N+j})||_2}{||A_c^{K_l}(\kappa_{N+j})||_2},$$

$$e_{\ell^\infty}(\kappa_{N+j}) = \frac{||A_c^{K_l}(\kappa_{N+j}) - A_c^{K_l,\text{pred}}(\kappa_{N+j})||_\infty}{||A_c^{K_l}(\kappa_{N+j})||_\infty},$$
(26)

$$e_F(\kappa_{N+j}) = \frac{||A_c^{K_l}(\kappa_{N+j}) - A_c^{K_l,\text{pred}}(\kappa_{N+j})||_F}{||A_c^{K_l}(\kappa_{N+j})||_F}.$$

A more important measure of the usefulness of the trained neural network is the predicted multiscale solution $u_{ms}^{\text{pred}}(\kappa)$ given by (23) and (24). We compared the predicted solution to u_{ms} defined by (11) and (12), and computed the relative error in L^2 and energy norm, i.e.,

$$
\begin{aligned}
e_{L^2}(\kappa_{N+j}) &= \left(\frac{\int_\Omega \left| u_{ms}(\kappa_{N+j}) - u_{ms}^{\text{pred}}(\kappa_{N+j}) \right|^2}{\int_\Omega \left| u_{ms}(\kappa_{N+j}) \right|^2} \right)^{\frac{1}{2}}, \\
e_a(\kappa_{N+j}) &= \left(\frac{\int_\Omega \kappa_j \left| \nabla u_{ms}(\kappa_{N+j}) - \nabla u_{ms}^{\text{pred}}(\kappa_{N+j}) \right|^2}{\int_\Omega \kappa_j \left| \nabla u_{ms}(\kappa_{N+j}) \right|^2} \right)^{\frac{1}{2}}.
\end{aligned}
\tag{27}
$$

4.1. Experiment 1

In this experiment, we considered curved channelized permeability fields. Each permeability field contained a straight channel and a curved channel. The straight channel was fixed and the curved channel struck the boundary of the target cell K_0 at the same points. The curvature of the sine-shaped channel inside K_0 varied among all realizations. We generated 2000 realizations of permeability fields, where the permeability coefficients were fixed. Samples of permeability fields are depicted in Figure 5. Among the 2000 realizations, 1980 sample pairs were randomly chosen and used as training samples, and the remaining 20 sample pairs were used as testing samples.

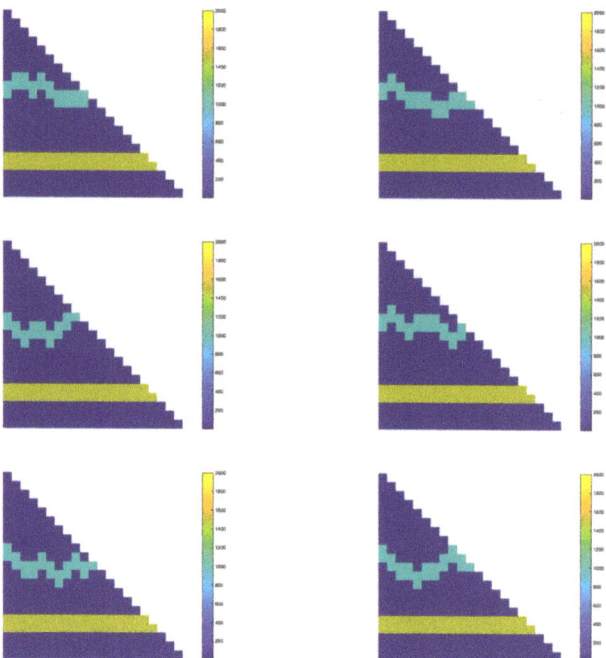

Figure 5. Samples of permeability fields in the target block K_0 in experiment 1.

For each realization, we computed the local multiscale basis functions and local coarse-scale stiffness matrix. In building the local snapshot space, we solved for harmonic extension of all the fine-grid boundary conditions. Local multiscale basis functions were then constructed by solving the spectral problem and multiplied the spectral basis functions with the multiscale partition of unity functions. With the offline space constructed, we computed the coarse-scale stiffness matrix. We used the training samples to build deep neural networks for approximating these GMsFEM quantities, and examined the performance of the approximations on the testing set.

Tables 1–3 record the error of the prediction by the neural networks in each testing sample and the mean error measured in the defined metric. It can be seen that the predictions were of high accuracy. This is vital in ensuring the predicted GMsFEM solver is useful. Table 4 records the error of the multiscale solution in each testing sample and the mean error using our proposed method. It can be observed that using the predicted GMsFEM solver, we obtained a good approximation of the multiscale solution compared with the exact GMsFEM solver.

Table 1. Percentage error of multiscale basis functions $\phi_1^{\omega_i}$ in experiment 1.

Sample	ω_1		ω_2		ω_3	
j	e_{L^2}	e_{H^1}	e_{L^2}	e_{H^1}	e_{L^2}	e_{H^1}
1	0.47%	3.2%	0.40%	3.6%	0.84%	5.1%
2	0.45%	4.4%	0.39%	3.3%	1.00%	6.3%
3	0.34%	2.3%	0.40%	3.1%	0.88%	4.3%
4	0.35%	4.2%	0.43%	5.4%	0.94%	6.6%
5	0.35%	3.3%	0.37%	3.9%	0.90%	6.1%
6	0.51%	4.7%	0.92%	12.0%	2.60%	19.0%
7	0.45%	4.1%	0.38%	3.2%	1.00%	6.4%
8	0.31%	3.4%	0.43%	5.5%	1.10%	7.7%
9	0.25%	2.2%	0.46%	5.6%	1.10%	6.2%
10	0.31%	3.5%	0.42%	4.5%	1.30%	7.6%
Mean	0.38%	3.5%	0.46%	5.0%	1.17%	7.5%

Table 2. Percentage error of multiscale basis functions $\phi_2^{\omega_i}$ in experiment 1.

Sample	ω_1		ω_2		ω_3	
j	e_{L^2}	e_{H^1}	e_{L^2}	e_{H^1}	e_{L^2}	e_{H^1}
1	0.47%	4.2%	0.40%	1.4%	0.32%	1.1%
2	0.57%	3.2%	0.31%	1.4%	0.30%	1.1%
3	0.58%	2.7%	0.31%	1.4%	0.33%	1.1%
4	0.59%	3.6%	0.13%	1.3%	0.32%	1.1%
5	0.53%	4.0%	0.51%	1.6%	0.27%	1.0%
6	0.85%	4.3%	0.51%	2.1%	0.29%	1.3%
7	0.50%	2.7%	0.22%	1.5%	0.29%	1.0%
8	0.43%	4.5%	0.61%	1.9%	0.35%	1.1%
9	0.71%	2.9%	0.14%	1.4%	0.27%	1.1%
10	0.66%	4.4%	0.53%	1.8%	0.26%	1.1%
Mean	0.59%	3.6%	0.37%	1.6%	0.30%	1.1%

Table 3. Percentage error of the local stiffness matrix $A_c^{K_0}$ in experiment 1.

Sample j	e_{ℓ^2}	e_F
1	0.67%	0.84%
2	0.37%	0.37%
3	0.32%	0.38%
4	1.32%	1.29%
5	0.51%	0.59%
6	4.43%	4.28%
7	0.34%	0.38%
8	0.86%	1.04%
9	1.00%	0.97%
10	0.90%	1.08%
Mean	0.76%	0.81%

Table 4. Percentage error of multiscale solution u_{ms} in experiment 1.

Sample j	e_{L^2}	e_a
1	0.31%	4.58%
2	0.30%	4.60%
3	0.30%	4.51%
4	0.27%	4.60%
5	0.29%	4.56%
6	0.47%	4.67%
7	0.39%	4.70%
8	0.30%	4.63%
9	0.35%	4.65%
10	0.31%	4.65%
Mean	0.33%	4.62%

4.2. Experiment 2

In this experiment, we considered sine-shaped channelized permeability fields. Each permeability field contained a straight channel and a sine-shaped channel. There were altogether five channel configurations, where the straight channel was fixed and the sine-shaped channel struck the boundary of the target cell K_0 at the same points. The curvature of the sine-shaped channel inside K_0 varied among these configurations. For each channel configuration, we generated 500 realizations of permeability fields, where the permeability coefficients followed random distributions. Samples of permeability fields are depicted in Figure 6. Among the 2500 realizations, 2475 sample pairs were randomly chosen and used as training samples, and the remaining 25 sample pairs were used as testing samples.

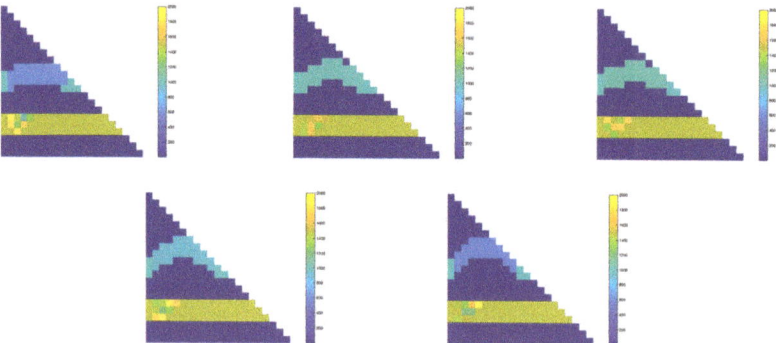

Figure 6. Samples of permeability fields in the target block K_0 in experiment 2.

Next, for each realization, we computed the local multiscale basis functions and local coarse-scale stiffness matrix. In building the local snapshot space, we solved for harmonic extension of randomized fine-grid boundary conditions, so as to reduce the number of local problems to be solved. Local multiscale basis functions were then constructed by solving the spectral problem and multiplied the spectral basis functions with the multiscale partition of unity functions. With the offline space constructed, we computed the coarse-scale stiffness matrix. We used the training samples to build deep neural networks for approximating these GMsFEM quantities, and examined the performance of the approximations on the testing set.

Figures 7–9 show the comparison of the multiscale basis functions in two respective coarse neighborhoods. It can be observed that the predicted multiscale basis functions were in good agreement with the exact ones. In particular, the neural network successfully interpreted the high conductivity regions as the support localization feature of the multiscale basis functions. Tables 5 and 6 record the mean error of the prediction by the neural networks, measured in the defined metric. Again, it can be seen that the prediction are of high accuracy. Table 7 records the mean error between the multiscale solution using the neural-network-based multiscale solver and using exact GMsFEM. we obtain a good approximation of the multiscale solution compared with the exact GMsFEM solver.

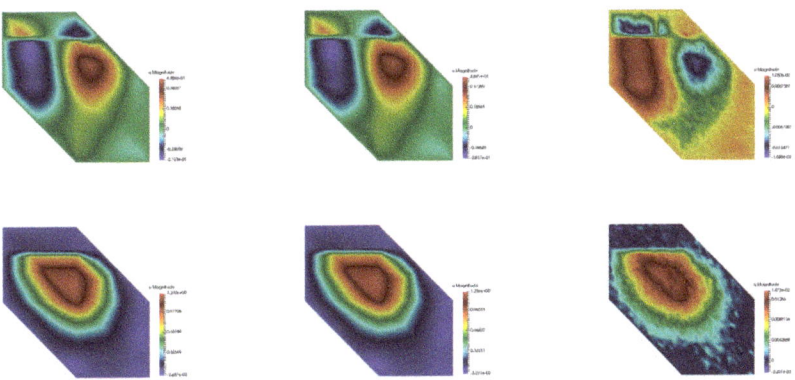

Figure 7. Exact multiscale basis functions $\phi_m^{\omega_1}$ (**left**), predicted multiscale basis functions $\phi_m^{\omega_1,\text{pred}}$ (**middle**) and their differences (**right**) in the coarse neighborhood ω_1 in experiment 2. The first row and the second row illustrate the first basis function $\phi_1^{\omega_1}$ and the second basis function $\phi_2^{\omega_1}$, respecitvely.

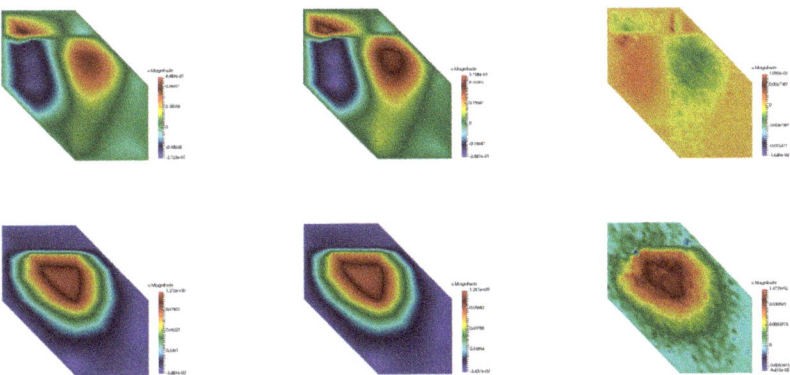

Figure 8. Exact multiscale basis functions $\phi_m^{\omega_2}$ (**left**), predicted multiscale basis functions $\phi_m^{\omega_2,\text{pred}}$ (**middle**) and their differences (**right**) in the coarse neighborhood ω_2 in experiment 2. The first row and the second row illustrate the first basis function $\phi_1^{\omega_2}$ and the second basis function $\phi_2^{\omega_2}$, respecitvely.

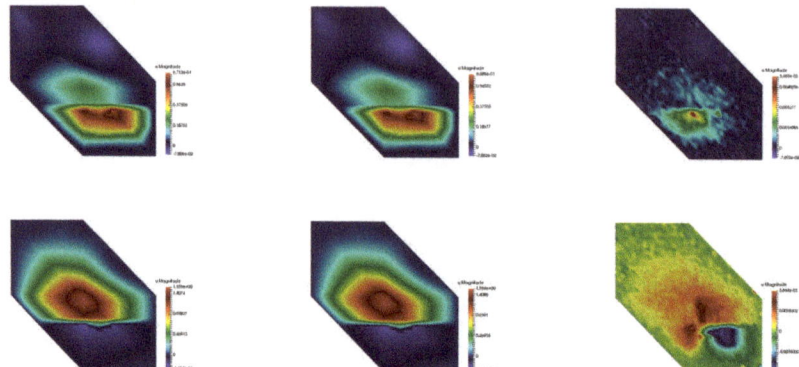

Figure 9. Exact multiscale basis functions $\phi_m^{\omega_3}$ (**left**), predicted multiscale basis functions $\phi_m^{\omega_3,\text{pred}}$ (**middle**) and their differences (**right**) in the coarse neighborhood ω_3 in experiment 2. The first row and the second row illustrate the first basis function $\phi_1^{\omega_3}$ and the second basis function $\phi_2^{\omega_3}$, respecitvely.

Table 5. Mean percentage error of multiscale basis functions $\phi_m^{\omega_i}$ in experiment 2.

Basis	ω_1		ω_2		ω_3	
m	e_{L^2}	e_{H^1}	e_{L^2}	e_{H^1}	e_{L^2}	e_{H^1}
1	0.55	0.91	0.37	3.02	0.20	0.63
2	0.80	1.48	2.17	3.55	0.27	1.51

Table 6. Percentage error of the local stiffness matrix $A_c^{K_0}$ in experiment 2.

	e_{ℓ^2}	e_{ℓ^∞}	e_F
Mean	0.75	0.72	0.80

Table 7. Percentage error of multiscale solution u_{ms} in experiment 2.

	e_{L^2}	e_a
Mean	0.03	0.26

5. Conclusions

In this paper, we develop a method using deep learning techniques for fast computation of GMsFEM discretizations. Given a particular permeability field, the main ingredients of GMsFEM, including the multiscale basis functions and coarse-scale matrices, are computed in an offline stage by solving local problems. However, when one is interested in calculating GMsFEM discretizations for multiple choices of permeability fields, repeatedly formulating and solving such local problems could become computationally expensive or even infeasible. Multi-layer networks are used to represent the nonlinear mapping from the fine-scale permeability field coefficients to the multiscale basis functions and the coarse-scale parameters. The networks provide a direct fast approximation of the GMsFEM ingredients in a local neighborhood for any online permeability fields, in contrast to repeatedly formulating and solving local problems. Numerical results are presented to show the performance of our proposed method. We see that, given sufficient samples of GMsFEM discretizations for supervised training, deep neural networks are capable of providing reasonably close approximations of the exact GMsFEM discretization. Moreover, the small consistency error provides good approximations of multiscale solutions.

Author Contributions: The authors have contributed equally to the work.

Funding: The research of Eric Chung is partially supported by the Hong Kong RGC General Research Fund (Project numbers 14304217 and 14302018) and CUHK Faculty of Science Direct Grant 2017-18. YE would like to acknowledge the support of Mega-grant of the Russian Federation Government (N 14.Y26.31.0013) and the partial support from NSF 1620318.

Conflicts of Interest: The authors declare no conflict of interest.

References

1. Efendiev, Y.; Hou, T. *Multiscale Finite Element Methods: Theory and Applications*; Surveys and Tutorials in the Applied Mathematical Sciences; Springer: New York, NY, USA, 2009; Volume 4.
2. Hou, T.; Wu, X. A multiscale finite element method for elliptic problems in composite materials and porous media. *J. Comput. Phys.* **1997**, *134*, 169–189. [CrossRef]
3. Jenny, P.; Lee, S.; Tchelepi, H. Multi-scale finite volume method for elliptic problems in subsurface flow simulation. *J. Comput. Phys.* **2003**, *187*, 47–67. [CrossRef]
4. Efendiev, Y.; Galvis, J.; Hou, T. Generalized multiscale finite element methods (GMsFEM). *J. Comput. Phys.* **2013**, *251*, 116–135. [CrossRef]
5. Chung, E.; Efendiev, Y.; Hou, T.Y. Adaptive multiscale model reduction with Generalized Multiscale Finite Element Methods. *J. Comput. Phys.* **2016**, *320*, 69–95. [CrossRef]
6. Efendiev, Y.; Leung, W.T.; Cheung, S.W.; Guha, N.; Hoang, V.H.; Mallick, B. Bayesian Multiscale Finite Element Methods. Modeling Missing Subgrid Information Probabilistically. *Int. J. Multiscale Comput. Eng.* **2017**. [CrossRef]
7. Cheung, S.W.; Guha, N. Dynamic Data-driven Bayesian GMsFEM. *arXiv* **2018**, arXiv:1806.05832.
8. Krizhevsky, A.; Sutskever, I.; Hinton, G.E. Imagenet classification with deep convolutional neural networks. *Adv. Neural Inf. Process. Syst.* **2012**, 1097–1105. Available online: http://www.informationweek.com/news/201202317 (accessed on 24 March 2019). [CrossRef]
9. Hinton, G.; Deng, L.; Yu, D.; Dahl, G.E.; Mohamed, A.R.; Jaitly, N.; Senior, A. Approximation Capabilities of Multilayer Feedforward Networks. *IEEE Signal Process. Mag.* **2012**, *29*, 82–97. [CrossRef]
10. He, K.; Zhang, X.; Ren, S.; Sun, J. Deep residual learning for image recognition. In Proceedings of the IEEE Conference on Computer Vision and Pattern Recognition, Las Vegas, NV, USA, 27–30 June 2016; pp. 770–778.
11. Cybenko, G. Approximations by superpositions of sigmoidal functions. *Math. Control Signals Syst.* **1989**, *2*, 303–314. [CrossRef]
12. Hornik, K. Approximation Capabilities of Multilayer Feedforward Networks. *Neural Netw.* **1991**, *4*, 251–257. [CrossRef]
13. Telgrasky, M. Benefits of depth in neural nets. *arXiv* **2016**, arXiv:1602.04485.
14. Liao, H.M.; Poggio, T. Learning functions: When is deep better than shallow. *arXiv* **2016**, arXiv:1603.00988v4.
15. Hanin, B. Universal function approximation by deep neural nets with bounded width and relu activations. *arXiv* **2017**, arXiv:1708.02691.
16. Zhu, Y.; Zabaras, N. Bayesian deep convolutional encoder–decoder networks for surrogate modeling and uncertainty quantification. *J. Comput. Phys.* **2018**, *366*, 415–447. [CrossRef]
17. Li, Z.; Shi, Z. Deep Residual Learning and PDEs on Manifold. *arXiv* **2017**, arXiv:1708.05115.
18. Weinan, E.; Yu, B. The Deep Ritz method: A deep learning-based numerical algorithm for solving variational problems. *arXiv* **2017**, arXiv:1710.00211.
19. Khoo, Y.; Lu, J.; Ying, L. Solving parametric PDE problems with artificial neural networks. *arXiv* **2017**, arXiv:1707.03351.
20. Cheung, S.W.; Chung, E.T.; Efendiev, Y.; Gildin, E.; Wang, Y. Deep Global Model Reduction Learning. *arXiv* **2018**, arXiv:1807.09335.
21. Wang, Y.; Cheung, S.W.; Chung, E.T.; Efendiev, Y.; Wang, M. Deep Multiscale Model Learning. *arXiv* **2018**, arXiv:1806.04830.
22. Chung, E.; Efendiev, Y.; Leung, W.T. Generalized multiscale finite element method for wave propagation in heterogeneous media. *SIAM Multicale Model. Simul.* **2014**, *12*, 1691–1721. [CrossRef]
23. Chung, E.; Efendiev, Y.; Lee, C. Mixed generalized multiscale finite element methods and applications. *SIAM Multicale Model. Simul.* **2014**, *13*, 338–366. [CrossRef]

24. Efendiev, Y.; Galvis, J.; Li, G.; Presho, M. Generalized Multiscale Finite Element Methods Oversampling strategies. *Int. J. Multiscale Comput. Eng.* **2013**, *12*, 465–484. [CrossRef]
25. Calo, V.; Efendiev, Y.; Galvis, J.; Li, G. Randomized Oversampling for Generalized Multiscale Finite Element Methods. *arXiv* **2014**, arXiv:1409.7114.
26. Chung, E.T.; Efendiev, Y.; Leung, W.T. Residual-driven online Generalized Multiscale Finite Element Methods. *J. Comput. Phys.* **2015**, *302*, 176–190. [CrossRef]
27. Chung, E.T.; Efendiev, Y.; Leung, W.T.; Vasilyeva, M.; Wang, Y. Non-local Multi-continua upscaling for flows in heterogeneous fractured media. *arXiv* **2018**, arXiv:1708.08379.
28. Glorot, X.; Bordes, A.; Bengio, Y. Deep Sparse Rectifier Neural Networks. In Proceedings of the Fourteenth International Conference on Artificial Intelligence and Statistics. PMLR, Ft. Lauderdale, FL, USA, 11–13 April 2011; pp. 315–323.
29. Kingma, D.P.; Ba, J. Adam: A method for stochastic optimization. *arXiv* **2014**, arXiv:1412.6980.
30. Chollet, F. Keras. 2015. Available online: https://keras.io (accessed on 1 May 2019).

© 2019 by the authors. Licensee MDPI, Basel, Switzerland. This article is an open access article distributed under the terms and conditions of the Creative Commons Attribution (CC BY) license (http://creativecommons.org/licenses/by/4.0/).

Review

Trigonometrically-Fitted Methods: A Review

Changbum Chun [1] and Beny Neta [2,*]

[1] Department of Mathematics, Sungkyunkwan University, Suwon 16419, Korea; cbchun@skku.edu
[2] Naval Postgraduate School, Department of Applied Mathematics, Monterey, CA 93943, USA
* Correspondence: bneta@nps.edu; Tel./Fax: +1-831-656-2235

Received: 28 October 2019; Accepted: 2 December 2019; Published: 6 December 2019

Abstract: Numerical methods for the solution of ordinary differential equations are based on polynomial interpolation. In 1952, Brock and Murray have suggested exponentials for the case that the solution is known to be of exponential type. In 1961, Gautschi came up with the idea of using information on the frequency of a solution to modify linear multistep methods by allowing the coefficients to depend on the frequency. Thus the methods integrate exactly appropriate trigonometric polynomials. This was done for both first order systems and second order initial value problems. Gautschi concluded that "the error reduction is not very substantial unless" the frequency estimate is close enough. As a result, no other work was done in this direction until 1984 when Neta and Ford showed that "Nyström's and Milne-Simpson's type methods for systems of first order initial value problems are not sensitive to changes in frequency". This opened the flood gates and since then there have been many papers on the subject.

Keywords: second order initial value problems; linear multistep methods; Obrechkoff schemes; trigonometrically fitted

1. Introduction

In this article, we discuss various methods for the numerical solution of the second order initial value problem

$$\begin{aligned} y'' &= f(x,y), \\ y(x_0) &= y_0, \\ y'(x_0) &= y'_0. \end{aligned} \qquad (1)$$

If the initial value problem contains $y'(x)$ then it is usually converted to a system of first order

$$\begin{aligned} y'_1(x) &= y_2, \\ y_1(x_0) &= y_0, \\ y'_2(x) &= f(x, y_1, y_2), \\ y_2(x_0) &= y'_0, \end{aligned} \qquad (2)$$

by defining new variables $y_1 = y, y_2 = y'$. In vector notation the system (2) can be written as

$$\mathbf{y}' = \mathbf{f}(x, \mathbf{y}), \ \mathbf{y}(x_0) = \mathbf{y}_0, \qquad (3)$$

where $\mathbf{y} = [y_1, y_2]^T$, $\mathbf{f} = [y_2, f(x, y_1, y_2)]^T$, and $\mathbf{y}_0 = [y_0, y'_0]^T$.

Here we are concerned with trigonometrically-fitted methods for (1) and (3).

There are several classes of methods, such as linear multistep methods (including Obrechkoff methods) and Runge-Kutta methods. Here we will introduce each class and then review the extension of those to solution of problems for which the frequency is approximately known in advance.

Linear multistep methods for the solution of (3) are given by

$$\sum_{j=0}^{k} \alpha_j y_{n+j} = h \sum_{j=0}^{k'} \beta_j f_{n+j}, \qquad (4)$$

and of (1) are given by

$$\sum_{j=0}^{k} \alpha_j y_{n+j} = h^2 \sum_{j=0}^{k'} \beta_j f_{n+j}, \qquad (5)$$

where y_{n+j} is the approximate value at x_{n+j} and similarly for f_{n+j}. In here k is called the *step-number* and k' is either $k-1$ or k. In the former case the method is called *explicit* and in the latter it is called *implicit*. The coefficients α_j and β_j are chosen to satisfy stability and convergence, as we describe in the sequel.

We now introduce the *first* and *second characteristic polynomials*,

$$\rho(\zeta) = \sum_{j=0}^{k} \alpha_j \zeta^j, \qquad (6)$$

$$\sigma(\zeta) = \sum_{j=0}^{k'} \beta_j \zeta^j. \qquad (7)$$

For (3) explicit methods for which $\rho(\zeta) = \zeta^k - \zeta^{k-1}$ are called *Adams-Bashforth* and the implicit ones are called *Adams-Moulton*. Explicit methods for which $\rho(\zeta) = \zeta^k - \zeta^{k-2}$ are called *Nyström methods* and the implicit ones are called *Generalized Milne-Simpson methods*. Gautschi [1] has developed Adams-type methods for first order equation as well as Nyström methods for the second order equation. Neta and Ford [2] only developed Nyström and Generalized Milne-Simpson methods for first order equation.

Definition 1. *If, for an arbitrary smooth enough test function $z(x)$, we have*

$$\sum_{j=0}^{k} \alpha_j z(x+jh) - h \sum_{j=0}^{k'} \beta_j z''(x+jh) = C_{p+1} h^{p+1} z^{(p+1)}(x) + O(h^{p+2}), \qquad (8)$$

then, p is called the order of the linear multistep method (4) and C_{p+1} is its error constant.

The expression given by (8) is called the local truncation error at x_{n+k} of the method (4), when $z(x)$ is the theoretical solution of the initial value problem (1).

In a similar fashion we have for (5)

$$\sum_{j=0}^{k} \alpha_j z(x+jh) - h^2 \sum_{j=0}^{k'} \beta_j z''(x+jh) = C_{p+2} h^{p+2} z^{(p+2)}(x) + O(h^{p+3}). \qquad (9)$$

Throughout, we shall assume that the linear multistep method (4) satisfies the following hypotheses (see [3]):

- $\alpha_k = 1, |\alpha_0| + |\beta_0| \neq 0, \sum_{j=0}^{k'} |\beta_j| \neq 0.$
- No common factors for the characteristic polynomials ρ and σ.
- $\rho(1) = 0, \rho'(1) = \sigma(1)$; this is a necessary and sufficient condition for the method to be *consistent*.
- The method is *zero-stable*; that is, all the roots ζ_ℓ of ρ satisfy $|\zeta_\ell| < 1$ for $\ell > 1$ and $\zeta_1 = 1$.

For the method (5) for second order initial value problems we have

- $\alpha_k = 1, |\alpha_0| + |\beta_0| \neq 0, \sum_{j=0}^{k'} |\beta_j| \neq 0.$
- No common factors for the characteristic polynomials ρ and σ.
- $\rho(1) = \rho'(1) = 0, \rho''(1) = 2\sigma(1)$; which is a necessary and sufficient condition for the method to be *consistent*.
- The method is *zero-stable*.

We now consider the test equation (see, e.g., Chawla and Neta [4])

$$y''(x) = -\lambda^2 y(x). \tag{10}$$

Let $\zeta_s, s = 1, 2, \ldots, k$ denote the zeros of the polynomial

$$\Omega(\zeta, H^2) = \rho(\zeta) + H^2 \sigma(\zeta), \tag{11}$$

for $H = \lambda h$ and let ζ_1, ζ_2 correspond to perturbations of the principal roots of $\rho(\zeta)$. We define *interval of periodicity* $(0, H^2)$ if, for all H^2 in the interval, the roots ζ_s of (11) satisfy $\zeta_1 = e^{i\theta(H)}, \zeta_2 = e^{-i\theta(H)}$, $|\zeta_s| \leq 1$, $s \geq 3$ and $\theta(H)$ is real.

If the interval of periodicity is $(0, \infty)$, then the method is called *P-stable*. Lambert and Watson [5] had shown that P-stable linear multistep methods are implicit of order at most 2.

Remark 1. *If the problem (1) has periodic solutions and the period is not known, then the P-stability is desirable. If the period is known approximately, then one can use the ideas in Gautschi [1], Neta and Ford [2], and others to be reviewed here.*

Another important property when solving (1) is the *phase lag* which was introduced by Brusa and Nigro [6]. Upon applying a linear two-step method to the test Equation (10), we obtain a difference equation of the form

$$A(H) y_{n+2} + B(H) y_{n+1} + C(H) y_n = 0, \tag{12}$$

whose solution is

$$y_n = B_1 \lambda_1^n + B_2 \lambda_2^n, \tag{13}$$

where B_1 and B_2 are constants depending on the initial conditions. The quadratic polynomial

$$A(H) \lambda^2 + B(H) \lambda + C(H) = 0, \tag{14}$$

is called the *stability polynomial*. The solutions to (14) are given by

$$\begin{aligned} \lambda_1 &= e^{(-a(H)+ib(H))H}, \\ \lambda_2 &= e^{(-a(H)-ib(H))H}. \end{aligned} \tag{15}$$

If $a(H) \equiv 0$ and $b(H) \equiv 1$, then we get the exact solution to the test Equation (10). The difference between the amplitudes of the exact solution of (10) and numerical solution is called *dissipation error*, see [7]. The expansion $b(H) - 1$ in powers of H is called *phase lag expansion*. The modulus of the leading terms is the *phase lag* of the method. See also Thomas [8] and Twizell [9].

Remark 2. *Raptis and Simos have developed methods with minimal phase-lag and also P-stable methods in [10–14].*

We now introduce an extension to the linear multistep methods. These are called *multiderivative* or *Obrechkoff* methods, see Obrechkoff [15] or [16].

For the first order equation we have

$$\sum_{j=0}^{k}\alpha_j y_{n-j+1} = \sum_{i=1}^{\ell}\sum_{j=0}^{k}\beta_{ij}h^i y_{n-j+1}^{(i)}, \qquad (16)$$

and for the second order equation

$$\sum_{j=0}^{k}\alpha_j y_{n-j+1} = \sum_{i=1}^{\ell}\sum_{j=0}^{k}\beta_{ij}h^{2i} y_{n-j+1}^{(2i)}. \qquad (17)$$

According to Lambert and Mitchell [17], the error constant decreases more rapidly with increasing ℓ rather than the step k. Thus, one can get one-step high order methods. A list of Obrechkoff methods for various k and ℓ is given in [17] for first order equations and in [18] for second order equations.

Several P-stable Obrechkoff methods for second-order initial-value problems (for $\ell \leq 3$) were derived by Van Daele and Vanden Berghe [19]. Ananthakrishnaiah [18] has also included the case $\ell = 4$.

Lastly, we introduce Runge-Kutta-Nyström (RKN) methods.

The general form of an explicit k-stage two-step Runge-Kutta-Nyström method (RKN) for the solution of (1) is given by, see Franco and Rández [20]

$$Y_i = (1+c_i)y_n - c_i y_{n-1} + h^2 \sum_{j=1}^{k} a_{ij} f\left(x_n + c_j h, Y_j\right), \qquad (18)$$

$$y_{n+1} = 2y_n - y_{n-1} + h^2 \sum_{i=1}^{k} b_i f\left(x_n + c_i h, Y_i\right). \qquad (19)$$

Vigo-Aguiar and Ramos [21] introduced methods based on Runge-Kutta collocation.

Definition 2. *Trigonometrically-fitted RKN method (18)–(19) integrates exactly the functions* $\sin(\lambda x)$ *and* $\cos(\lambda x)$ *with* $\lambda > 0$ *the principal frequency of the problem when applied to the test Equation (10).*

In general, a method integrates exactly the set of functions $\{u_1(x), u_2(x), \ldots, u_r(x)\}$, $r \leq k$ if the following conditions are satisfied

$$u_\ell(x_n + h) = 2u_\ell(x_n) - u_\ell(x_n - h) + h^2 \sum_{i=1}^{k} b_i u_\ell''(x_n + c_i h), \; \ell = 1, \ldots, r$$

$$u_\ell(x_n + c_i h) = (1+c_i)u_\ell(x_n) - c_i u_\ell(x_n - h) + h^2 \sum_{j=1}^{k} a_{ij} u_\ell''(x_n + c_j h), \qquad (20)$$

$$i = 1, \ldots, k, \, \ell = 1, \ldots, r$$

2. Methods Based on Linear Multistep Methods

The idea of fitting functions other than monomials goes back to Greenwood [22], Brock and Murray [23], Dennis [24], Gautschi [1] and Salzer [25].

The first paper suggesting the use of the frequency of the solution is due to Gautschi [1]. He considered Störmer type methods for the solution of (1). The idea is to allow the coefficients to depend on the frequency ω. Let $\mathcal{L}$ be a functional defined by

$$\mathcal{L}y = \sum_{j=0}^{k}\left[\alpha_j y(x_0 + (n+1-j)h) - h\beta_j f(x_0 + (n+1-j)h)\right], \qquad (21)$$

where $\alpha_0 = 1$. Since we are introducing trigonometric functions, we refer to order as *algebraic order* and define *trigonometric order* as follows:

Definition 3. *A linear functional* $\mathcal{L} \in C^s[a,b]$ *is said to be of algebraic order p, if*

$$\mathcal{L}x^r \equiv 0, \ r = 0, 1, \ldots, p,$$

and $\mathcal{L}x^{p+1}$ *does not vanish. Therefore we have* $p+1$ *conditions for methods of algebraic order p.*

The method

$$\sum_{j=0}^{k} \alpha_j y_{n+j} = h \sum_{j=0}^{k'} \beta_j(v) f_{n+j}, \tag{22}$$

where $v = \omega h$ *and* $\alpha_k = 1$ *is said to be of trigonometric order q relative to the frequency* ω *if the associated linear functional*

$$\mathcal{L}y(x) = \sum_{j=0}^{k} \alpha_j y_{n+j} - h \sum_{j=0}^{k'} \beta_j(v) y'_{n+j}$$

satisfies

$$\mathcal{L}1 \equiv 0,$$

and

$$\mathcal{L}\cos(r\omega x) \equiv \mathcal{L}\sin(r\omega x) \equiv 0, \ r = 1, 2, \ldots, q,$$

and $\mathcal{L}\cos((q+1)\omega x)$ *and* $\mathcal{L}\sin((q+1)\omega x)$ *are not both identically zero.*

Therefore, methods of trigonometric order q satisfy $2q+1$ conditions.

Linear multistep or trigonometrically fitted method for second order ordinary differential equations (ODEs) (1) satisfy an additional condition

$$\mathcal{L}x \equiv 0$$

for the same order, see Lambert [3].

Remark 3. *The trigonometric order is lower than the algebraic order, since the trigonometric polynomials requires two conditions for each degree, see Lambert [3].*

Gautschi [1] allowed the coefficients α_j to depend on v and listed several explicit and implicit methods of trigonometric orders $q \leq 3$. The form of the explicit methods is:

$$y_{n+1} + \alpha_{q1}(v)y_n + \alpha_{q2}(v)y_{n-1} = h^2 \sum_{j=1}^{2q-1} \beta_{qj}(v) f_{n+1-j}. \tag{23}$$

We only list the methods of trigonometric orders 1 and 2 using powers of $\cos(v)$ instead of the Taylor series expansions shown in [1]. Those Taylor series expansions should be used when $h \to 0$.

For $q = 1$, the coefficients are:

$$\alpha_{11} = -2, \ \alpha_{12} = 1, \ \beta_{11} = \left(\frac{2\sin(v/2)}{v}\right)^2. \tag{24}$$

For $q = 2$, the coefficients are:

$$\alpha_{21} = \frac{2}{3}(\cos(2v) - 4\cos(v)),$$

$$\alpha_{22} = -\alpha_{21} - 1,$$

$$\beta_{21} = \frac{1}{6}\frac{-16\cos(v)^3 + 9\cos(v) + 7}{v^2(2\cos(v)+1)},$$

$$\beta_{22} = \frac{1}{3}\frac{8\cos(v)^3 - 9\cos(v)^2 - 3\cos(v) + 4}{v^2(2\cos(v)+1)},$$

$$\beta_{23} = \frac{1}{2}\frac{1-\cos(v)}{v^2(2\cos(v)+1)}.$$

(25)

The form of the implicit methods is:

$$y_{n+1} + \alpha_{q1}(v)y_n + \alpha_{q2}(v)y_{n-1} = h^2 \sum_{j=0}^{2q-2} \beta_{qj}(v)f_{n+1-j}.$$

(26)

For $q = 1$, the coefficients are:

$$\alpha_{11} = \frac{2\cos(v)}{1 - 2\cos(v)},$$

$$\alpha_{12} = -\alpha_{11} - 1,$$

$$\beta_{10} = \frac{2(1 - \cos(v))}{v^2(2\cos(v) - 1)}.$$

(27)

For $q = 2$, the coefficients are:

$$\alpha_{21} = -2,$$

$$\alpha_{22} = 1,$$

$$\beta_{20} = \frac{1}{2}\frac{1-\cos(v)}{v^2(2\cos(v)+1)},$$

$$\beta_{21} = \frac{2+\cos(v)-3\cos(v)^2}{v^2(2\cos(v)+1)},$$

$$\beta_{22} = \beta_{20}.$$

(28)

Neta and Ford [2] have constructed the *Nyström* and *Generalized Milne-Simpson methods* for a first order (3) where the coefficients β_j are functions of the frequency. Here we list a few of those.

For $q = 1$, the explicit method is

$$y_{n+2} - y_n = \frac{2\sin(v)}{v}hf_{n+1}.$$

(29)

For $q = 2$, the explicit method is

$$y_{n+4} - y_{n+2} = -h\frac{\sin(v)}{v(1+2\cos(v))}[f_n + 2(1-2\cos(v))(1+\cos(v))f_{n+1} \quad (30)$$

$$+(4\cos(v)\cos(2v)+1)f_{n+2} - 4\cos(v)(1+\cos(v))f_{n+3}].$$

For $q = 1$, the implicit method is a one-parameter family

$$y_{n+2} - y_n = h\left[\beta_0 f_n + \left(\frac{2\sin(v)}{v} - 2\beta_0\cos(v)\right)f_{n+1} + \beta_0 f_{n+2}\right]. \quad (31)$$

Note that the choice $\beta_0 = 0$ leads to the explicit method (29).
For $q = 2$, the implicit method is

$$y_{n+3} - y_{n+1} = h\frac{\sin(v)}{v(1+2\cos(v))}[f_{n+1} + 2(1+\cos(v))f_{n+2} + f_{n+3}]. \quad (32)$$

Vigo-Aguiar and Ramos [26] show how to choose the frequency for nonlinear ODEs. Van der Houwen and Sommeijer [27] have developed predictor-corrector methods. Neta [28] has developed exponentially fitted methods for problems whose oscillatory solution is damped. Raptis and Allison [29] have used the idea for the solution of Schrödinger equation. Stiefel and Bettis [30] have stabilized Cowell's method [31] by fitting trigonometric polynomials. Lambert and Watson [5] introduced symmetric multistep methods which have non-vanishing interval of periodicity. Quinlan and Termaine [32] have developed high order symmetric multistep methods. Simos and Vigo-Aguiar [33] have developed exponentially-fitted symmetric methods of algebraic order eight based on the work of [32]. They demonstrated the superiority of their method on two orbital examples integrated on a long time interval $t \in [0, 10^5]$. Another idea developed by Neta and Lipowski [34] is to use the energy of the system instead of integrating the angular velocity. They have demonstrated the benefit of their method using several examples for perturbation-free flight and a more general case on long time flight. Vigo-Aguiar and Ferrándiz [35] have developed a general procedure for the adaptation of multistep methods to numerically solve problems having periodic solutions. Vigo-Aguiar et al. [36] and Martín-Vaquero and Vigo-Aguiar [37] have developed methods for stiff problems by using Backward Differentiation Formulae (BDF) methods. See also Neta [38].

Sommeijer et al. [39] have suggested a different idea for trigonometrically-fitted methods. Instead of requiring fitting cosine and sine functions of multiple of the frequency, they suggest taking several frequencies in some interval around the known frequency. The frequencies are chosen to be the roots of a Chebyshev polynomial, so that we minimize the maximum error. Such methods were called minimax methods. See also Neta [40].

We now give more details. Suppose we have an interval $[\underline{\omega}, \bar{\omega}]$ of frequencies and we pick q frequencies

$$\omega_j = \frac{1}{2}\left((\bar{\omega})^2 + (\underline{\omega})^2\right) + \frac{1}{2}\left((\bar{\omega})^2 - (\underline{\omega})^2\right)\cos(\frac{2j-1}{2q}\pi)]^{1/2}.$$

The idea is to interpolate the sine and cosine functions of those frequencies

$$\mathcal{L}1 \equiv 0,$$

and

$$\mathcal{L}\cos(\omega_r x) \equiv \mathcal{L}\sin(\omega_r x) \equiv 0, \quad r = 1, 2, \ldots, q.$$

Thus for the second order initial value problem, we have the system

$$(h\omega_j)^2 \left\{ \sum_{\ell=0}^{k/2-1} 2b_\ell \cos((k/2-\ell)h\omega_j) + b_{k/2} \right\} = -\sum_{\ell=0}^{k} a_\ell \cos((k/2-\ell)h\omega_j),$$

for $j = 1, \ldots, q$. Unfortunately, this yields very messy coefficients and we will not list any of them here.

3. Methods Based on Obrechkoff Methods

Simos [41] has developed a P-stable trigonometrically-fitted Obrechkoff method of algebraic order 10 for (1).

$$y_{n+1} - 2y_n + y_{n-1} = \sum_{j=1}^{3} h^{2j} \left[b_{j0} y_{n+1}^{(2j)} + 2b_{j1} y_n^{(2j)} + b_{j0} y_{n-1}^{(2j)} \right], \tag{33}$$

where

$$b_{10} = \frac{89}{1878} - \frac{15120}{313} b_{31},$$

$$b_{11} = \frac{425}{939} + \frac{15120}{313} b_{31},$$

$$b_{20} = -\frac{1907}{1577520} + \frac{660}{313} b_{31}, \tag{34}$$

$$b_{21} = \frac{30257}{1577520} + \frac{690}{313} b_{31},$$

$$b_{30} = \frac{59}{3155040} - \frac{13}{313} b_{31}.$$

In order to ensure P-stability, the coefficient b_{31} must be

$$b_{31} = \Big(190816819200[1 - \cos(H)] - 95408409600H^2 + 7950700800H^4$$
$$- 265023360H^6 + 4732560H^8 - 52584H^{10} + 1727H^{12} \Big) / (3568320H^{12}). \tag{35}$$

The method requires an approximation of the first derivative which is given by

$$y'_{n+1} = \frac{1}{2h}(y_{n-1} - 4y_n + 3y_{n+1}) - \frac{h}{12}(y''_{n-1} + 2y''_n + 3y''_{n+1}). \tag{36}$$

He showed that the local truncation error is

$$LTE = \left(-\frac{2923}{209898501120} + \frac{59}{1577520} b_{31} \right) h^{12} y_n^{(12)}.$$

Wang et al. [42] have suggested a slight modification to the coefficient b_{31} as follows

$$b_{31} = \frac{3155040 - 1428000H^2 + 60514H^4 - a_1 \cos(H)}{5040H^2(-15120 + 6900H^2 - 313H^4 + a_2 \cos(H))}, \tag{37}$$

where $a_1 = 3155040 + 149520H^2 + 3814H^4 + 59H^6$ and $a_2 = 15120 + 660H^2 + 13H^4$.

Wang et al. [42] have developed a method of algebraic order 12 as follows

$$\begin{aligned}y_{n+1} - 2y_n + y_{n-1} &= h^2\left(\alpha_1\left(y''_{n+1} + y''_{n-1}\right) + \alpha_2 y''_n\right) \\ &+ h^4\left(\beta_1\left(y''_{n+1} + y''_{n-1}\right) + \beta_2 y''_n\right) \\ &+ h^6\left(\gamma_1\left(y''_{n+1} + y''_{n-1}\right) + \gamma_2 y''_n\right),\end{aligned} \quad (38)$$

where

$$\alpha_1 = \frac{229}{7788}, \quad \beta_1 = -\frac{1}{2360}, \quad \beta_2 = \frac{711}{12980},$$

$$\gamma_1 = \frac{127}{39251520}, \quad \gamma_2 = \frac{2923}{3925152},$$

and α_2 is chosen so the method is P-stable,

$$\alpha_2 = -2H^{-2} + H^2\beta_2 - H^4\gamma_2 + 2\cos(H)\left(H^{-2} - \alpha_1 + H^2\beta_1 - H^4\gamma_1\right).$$

The method is of algebraic order 12 and the local truncation error is now

$$LTE = \frac{45469}{1697361329664000} h^{14}\left(\omega^{12} y''_n - y_n^{(14)}\right).$$

The first order derivative is obtained by

$$y'_{n+1} = \frac{1}{66h}\left(305 y_{n+1} - 544 y_n + 239 y_{n-1}\right) + \frac{h}{1980}\left(-5728 y''_n - 571 y''_{n-1} + 119 y''_{n+1}\right)$$

$$+ \frac{h^2}{2970}\left(128 y'''_n - 173 y'''_{n-1}\right) + \frac{h^3}{2970}\left(-346 y^{(4)}_n - 13 y^{(4)}_{n-1}\right) + \frac{h^5}{62370}\left(-71 y^{(6)}_n + y^{(6)}_{n-1}\right).$$

Remark 4. *Neta [43] has developed a P-stable method of algebraic order 18.*

Vanden Berghe and Van Daele [44] have suggested fitting a combination of monomials and exponentials, i.e., the set $\{1, x, \ldots, x^K, e^{\pm\mu x}, xe^{\pm\mu x}, \ldots, x^P e^{\pm\mu x}\}$. Clearly when μ is purely imaginary, we get the cosine and sine functions. When $K = -1$, we get only the exponential functions and when $P = -1$ we get only monomials (which is the well known Obrechkoff method). Even when $K = -1$, we are not getting the cosine and sine functions of multiples of the frequency as in the previously discussed methods. They developed methods of algebraic order 8. Here we list only two of those, one with $K = 5, P = 1$ (39) and the other with $K = 7, P = 0$ (40).

The first method is given by

$$b_{10} = \frac{1}{12} - 2b_{20} - 2b_{21},$$

$$b_{11} = \frac{5}{12} + 2b_{20} + 2b_{21},$$

$$b_{20} = \frac{v^5 \sin(v) + 2(\cos(v) + 5)v^4 + 48(\cos(v) - 1)A}{12v^4(v^3 \sin(v) - 4(1 - \cos(v))^2)},$$

$$b_{21} = \frac{5v^5 \sin(v) - 2\cos(v)(\cos(v) + 5)v^4 - 48(\cos(v) - 1)B}{12v^4(v^3 \sin(v) - 4(1 - \cos(v))^2)},$$

(39)

where $A = (v^2 + \cos(v) - 1)$ and $B = (v^2 \cos(v) + \cos(v) - 1)$.

The second method is

$$b_{10} = \frac{1}{30} - 12b_{20},$$

$$b_{11} = \frac{7}{15} + 12b_{20},$$

$$b_{20} = \frac{4\cos(v)v^2 + 56v^2 - 3v^4 + 120\cos(v) - 120}{120v^2(12\cos(v) - 12 + \cos(v)v^2 + 5v^2)},$$

$$b_{21} = \frac{1}{40} + 5b_{20}.$$

(40)

4. Methods Based on Runge-Kutta

For a trigonometrically-fitted method, we have (see Franco and Rández [20])

$$\sum_{i=1}^{k} b_i \cos(c_i v) = \frac{2(\cos(v) - 1)}{v^2},$$

$$\sum_{i=1}^{k} b_i \sin(c_i v) = 0,$$

$$\sum_{j=1}^{k} a_{ij} \cos(c_i v) = \frac{\cos(c_i v) + c_i \cos(v) - (1 + c_i)}{v^2}, \; i = 1, \ldots, k,$$

$$\sum_{j=1}^{k} a_{ij} \sin(c_i v) = \frac{\sin(c_i v) - c_i \sin(v)}{v^2}, \; i = 1, \ldots, k.$$

(41)

The solution for $k = 3$ is given in Franco [45]

$$\begin{aligned}
c_1 &= -1, \\
c_2 &= 0, \\
c_3 &= 1, \\
b_1 &= \frac{2\cos(v) - 2 - v^2}{2v^2(\cos(v) - 1)}, \\
b_2 &= \frac{(2 - v^2)\cos(v) - 2}{v^2(\cos(v) - 1)}, \\
b_3 &= b_1, \\
a_{31} &= 0, \\
a_{32} &= \frac{2(\cos(v) - 1)}{v^2}, \\
a_{33} &= 0.
\end{aligned}$$

(42)

This method has an algebraic order 2 and reduces to the two-stage explicit Numerov method of Chawla [46]. The method integrates exactly the set of functions $\{1, x, x^2, \cos(\omega x), \sin(\omega x)\}$, similar to the idea of Vanden Berghe and Van Daele [44].

Franco and Rández [20] have developed a 7-stage method of algebraic order 7 which integrates exacly the monomials up to x^6 and $\sin(\omega x)$ and $\cos(\omega x)$. A 5-stage family of methods of algebraic order 6 listed here and in Tsitouras [47] has been developed by Fang et al. [48]. Here we list just one member of the family.

$$c_1 = -1, \; c_2 = 0, \; c_3 = \frac{1}{2}, \; c_4 = -\frac{1}{2}, \; c_5 = 1,$$

(43)

$$a_{31} = -\frac{\sin^2(v/4)}{v^2\cos(v/2)},$$
$$a_{32} = \frac{2\cos(v/2) + 2\cot(v)\sin(v/2) - 3}{2v^2},$$
$$a_{41} = \frac{36 + v^2 - 36\cos(v/2)}{72v^2\cos(v/2)},$$
$$a_{42} = \frac{36\sin(v/2) - v^2\sin(3v/2) - 18\sin(v)}{36v^2\sin(v)},$$
$$a_{43} = \frac{1}{36}, \tag{44}$$
$$a_{51} = -\frac{2}{9\cos(v/2)},$$
$$a_{52} = \frac{2\cos(v) - 2}{v^2} - \frac{1}{3\cos(v/2)} - \frac{2\sin(3v/2)}{9\sin(v)},$$
$$a_{53} = \frac{2}{9},$$
$$a_{54} = \frac{2}{3},$$
$$b_1 = b_5 = \frac{6\cos(v) - 6 - v^2 - 2v^2\cos(v/2)}{48v^2\sin^4(v/4)},$$
$$b_2 = \frac{(18 + v^2)\cos(v) - 18 - 10v^2\cos(v/2)}{24v^2\sin^4(v/4)}, \tag{45}$$
$$b_3 = b_4 = \frac{12 + 5v^2 + (v^2 - 12)\cos(v)}{24v^2\sin^4(v/4)}.$$

Demba et al. [49] have developed fourth and fifth order Runge-Kutta-Nyström trigonometrically-fitted methods for (1). The idea is based on using 3-stage method to get a 4-stage trigonometrically-fitted method. Here we list the coefficients

$$y_{n+1} = y_n + hy'_n + h^2\sum_{i=1}^{s} b_i f(x_n + hc_i, Y_i) \tag{46}$$

$$y'_{n+1} = y'_n + h\sum_{i=1}^{s} d_i f(x_n + hc_i, Y_i) \tag{47}$$

where Y_i are given by (18) and

$$c_1 = 0,$$
$$c_2 = \frac{3}{16}\frac{v^3\cos(v) - 5v^2\sin(v) + 4v^3 - 32v\cos(v) + 160\sin(v) - 128v}{v(6v\sin(v) + v^2 + 30\cos(v) - 30)},$$
$$c_3 = -\frac{3}{500}\frac{-11v^6(4 + \cos(v)) + 55v^5\sin(v) + v^4(38\cos(v) + 1536) + T_1 + T_2}{v^2(6v\sin(v) + v^2 + 30\cos(v) - 30)},$$
$$a_{2,1} = \frac{1}{32}, \tag{48}$$
$$a_{3,1} = -\frac{1}{1000}\frac{-11v^5 + 384v^3 + 2112\sin(v) - 2112v}{v^3},$$
$$a_{3,2} = \frac{44}{125},$$

where

$$T_1 = -1920v^3\sin(v) + 2112v\cos(v)\sin(v) - v^2(672\cos(v) + 4448) + 10560\cos(v)^2,$$

and

$$T_2 = +3008v\sin(v) - 21120\cos(v) + 10560,$$

$$b_1 = \frac{1}{24},$$
$$b_2 = \frac{16}{165} \frac{v^4 + 66v\sin(v) - 21v^2 + 330\cos(v) - 330}{v^2(v^2 - 32)},$$
$$b_3 = \frac{25}{264},$$
$$d_1 = \frac{1}{24}, \qquad (49)$$
$$d_2 = \frac{16}{33},$$
$$d_3 = \frac{125}{264}.$$

The Taylor series expansion of the coefficients is given by

$$\begin{aligned}
a_{3,1} &= -4/125 - (33/5000)v^2 + (11/26250)v^4 - (11/1890000)v^6,\\
b_2 &= 4/11 - (1/3600)v^4 + (1/161280)v^6,\\
c_2 &= 1/4 - (1/26880)v^4 + (19/7741440)v^6,\\
c_3 &= 4/5 + (13/4375)v^4 - (257/3780000)v^6.
\end{aligned} \qquad (50)$$

5. Comments on Order

Definition 1 of order (see (8) and (9)) can be extended to trigonometrically fitted methods. Note that a method is of order p for first (second) order ODEs if it fits monomomials up to degree $p+1$ ($p+2$), respectively. Therefore, methods of trigonometric order q are methods of order $2q$. Method, such as (39) for second order ODEs is of eighth order, since it fits monomial up to degree 5, and $x^n \cos(\omega x)$, $x^n \sin(\omega x)$, $n = 0, 1$. In Table 1, we will list all methods used in the examples with their order.

Table 1. The order of methods used in the examples for first and second order ODEs.

Method	First Order ODEs	Second Order ODEs
(25)		4
(30)	4	
(51)		4
AI2	4	
AI3	6	
(32)	4	
(66)	6	
(46)	4	

6. Numerical Examples

The methods developed originally by Gautschi [1] and those that follow by Neta and Ford [2] fit low order monomials and the sine and cosine functions of multiples of the frequency. On the other hand, the methods developed later by Vanden Berghe and Van Daele [44] use monomials and product of monomials and sine and cosine functions of the frequency. We will demonstrate via the first three examples the difference between the two strategies. Vanden Berghe and Van Daele [50] compared the two approaches in some cases but not used schemes developed by Neta and Ford [2]. In the latter examples we also compare the results to Runge-Kutta-Nyström based method (46)–(49), see [49].

First, we list a method of trigonometric order 2 based on the idea of Vanden Berghe and Van Daele [44], which we obtained using Maple software, see Chun and Neta [51].

$$y_{n+1} + a_1 y_n + a_2 y_{n-1} = h^2(b_1 f_n + b_2 f_{n-1} + b_3 f_{n-2}), \qquad (51)$$

where

$$\begin{aligned}
a_1 &= -\sin(v)v - 2\cos(v), \\
a_2 &= -1 - a_1, \\
b_1 &= \frac{v(v\sin(v) - 1)(\cos(v) + 1) + 2\sin(v)}{v^3(1 + \cos(v))}, \\
b_2 &= \frac{v(2 - v\sin(v))(\cos(v) + 1) - 4\sin(v)\cos(v)}{v^3(1 + \cos(v))}, \\
b_3 &= \frac{2 - v\sin(v) - 2\cos(v)}{v^3 \sin(v)}.
\end{aligned} \qquad (52)$$

The Taylor series expansion of the coefficients are

$$\begin{aligned}
a_1 &= -2 + \frac{1}{12}v^4 - \frac{1}{180}v^6, \\
a_2 &= -1 - a_1, \\
b_1 &= \frac{13}{12} - \frac{19}{120}v^2 + \frac{37}{4032}v^4 - \frac{41}{362880}v^6, \\
b_2 &= -\frac{1}{6} + \frac{3}{20}v^2 - \frac{59}{10080}v^4 + \frac{13}{36288}v^6, \\
b_3 &= \frac{1}{12} + \frac{1}{120}v^2 + \frac{17}{20160}v^4 + \frac{31}{362880}v^6.
\end{aligned} \qquad (53)$$

Example 1. *The first example is chosen so that the exact solution in a combination of sine and cosine of multiples of the frequency, i.e.,*

$$y''(x) + 9y(x) = 3\sin(6x), \quad 0 \le x \le 40\pi \qquad (54)$$

subject to the initial conditions

$$y(0) = 1, \ y'(0) = 3. \qquad (55)$$

The exact solution is

$$y_{exact}(x) = \frac{11}{9}\sin(3x) + \cos(3x) - \frac{1}{9}\sin(6x). \qquad (56)$$

The results using $h = \pi/500$ are given in Table 2. We expect the methods that fit sine and cosine of multiples of the frequency will do better.

Table 2. The L_2 norm of the error for the first example using three methods for three different values around the exact frequency.

Method	ω	L_2 Error
(25)	2.95	0.984529(-5)
(25)	3.	0.215491(-43)
(25)	3.05	0.109415(-4)
(30)	2.95	0.155534(-6)
(30)	3	0.974885(-16)
(30)	3.05	0.168859(-6)
(51)	2.95	0.113080(-6)
(51)	3.	0.396444(-9)
(51)	3.05	0.116944(-6)

153

Based on the results we see that (25) is best when the frequency is known exactly. If it is not known exactly, the method prefers underestimation of the frequency. The second best is (30). This method will have no preference to underestimation.

Example 2. *The second example is very similar*

$$y''(x) + 9y(x) = 3\sin(3x), \ 0 \le x \le 40\pi \tag{57}$$

subject to the initial conditions

$$y(0) = 1, \ y'(0) = 3. \tag{58}$$

The exact solution is

$$y_{exact}(x) = \frac{7}{6}\sin(3x) + \cos(3x) - \frac{1}{2}x\cos(3x). \tag{59}$$

The results are given in Table 3. Now we expect that the method (51) due to Chun and Neta [51] will perform better, since the exact solution has a product of monomial and cosine. In fact this is the case followed by (25). The method (30) requires smaller step size to converge and the results are not as good as those of the other two methods. Note that for this example all methods have no preference to underestimation of the frequency.

Table 3. The L_2 norm of the error for the second example using three methods for three different values around the exact frequency.

Method	ω	L_2 Error
(25)	2.95	0.302359(-3)
(25)	3.	0.109032(-5)
(25)	3.05	0.338322(-3)
(30)	2.95	0.448371(-2)
(30)	3	0.447345(-2)
(30)	3.05	0.446280(-2)
(51)	2.95	0.347311(-5)
(51)	3.	0.134979(-40)
(51)	3.05	0.364011(-5)

Example 3. *What if the frequency of the forcing term is not an integer multiple of the frequency of the homogeneous solution? We now consider the following example*

$$y''(x) + 9y(x) = 3\sin(4x), \ 0 \le x \le 40\pi \tag{60}$$

subject to the initial conditions

$$y(0) = 1, \ y'(0) = 3. \tag{61}$$

The exact solution is

$$y_{exact}(x) = \frac{11}{7}\sin(3x) + \cos(3x) - \frac{3}{7}\sin(4x). \tag{62}$$

The results are given in Table 4. Now (51) is best followed by (30).

Table 4. The L_2 norm of the error for the third example using three methods for three different values around the exact frequency.

Method	ω	L_2 Error
(25)	2.95	0.980898(-5)
(25)	3.	0.195799(-9)
(25)	3.05	0.109015(-4)
(30)	2.95	0.286524(-6)
(30)	3	0.412154(-6)
(30)	3.05	0.557602(-6)
(51)	2.95	0.112995(-6)
(51)	3.	0.685320(-10)
(51)	3.05	0.116839(-6)

Based on the three examples, we find that (51) is best in the last two examples, but not in the first case where the frequency of the forcing term is a multiple of the frequency of the homogeneous solution.

Before moving to the rest of the experiments, we have decided to rerun the first example on a much longer interval. This will test the quality of those methods in long-term integration. We have taken the same step size $h = \pi/500$ and integrated for $0 \leq x \leq 4000\pi$. The results are given in Table 5. It is clear that the method due to Neta and Ford is no longer viable. The method (51) gave same errors when $\omega = 3$ but all other cases show lower accuracy at the end of the longer interval

Table 5. The L_2 norm of the error for the first example using three methods for three different values around the exact frequency.

Method	ω	L_2 Error
(25)	2.95	0.985005(-3)
(25)	3.	0.102448(-41)
(25)	3.05	0.109355(-2)
(30)	2.95	Div.
(30)	3	Div.
(30)	3.05	Div.
(51)	2.95	0.113480(-4)
(51)	3.	0.396444(-9)
(51)	3.05	0.117327(-4)

Example 4. *The fourth example is a system of two second order initial value problems*

$$u''(x) = -\frac{u(x)}{(u^2(x) + v^2(x))^{3/2}}, \quad 0 \leq x \leq 12\pi \tag{63}$$

$$v''(x) = -\frac{v(x)}{(u^2(x) + v^2(x))^{3/2}}, \quad 0 \leq x \leq 12\pi$$

subject to the initial conditions

$$u(0) = 0, \; u'(0) = 1,$$
$$v(0) = 1, \; v'(0) = 0. \tag{64}$$

The exact solution is given by

$$u_{exact}(x) = \sin(x), \; v_{exact}(x) = \cos(x). \tag{65}$$

We have converted this to a system of first order equations and solved it numerically using implicit Adams methods of trigonometric orders 2 and 3 (denoted AI2 and AI3, respectively) and generalized Milne-Simpson methods (GMS) of the same order, which are (32) and (66), respectively. We also included results from [51] and Runge-Kutta-Nyström method [49]. The results are given in Table 6. For Adams implicit, we have used the Taylor series coefficients as given in [1]. For GMS with $q = 2$, we used the coefficients as given in [2]. They did not give the coefficients for $q = 3$ but suggested to numerically solve the system for the coefficients. We were able now to get the coefficients

$$y_{n+5} = y_{n+3} + h\left(b_0 f_n + b_1 f_{n+1} + b_2 f_{n+2} + b_3 f_{n+3} + b_4 f_{n+4} + b_5 f_{n+5}\right), \tag{66}$$

where

$$\begin{aligned}
b_0 &= \frac{1}{6}\frac{\sin(v)}{d_1}, \\
b_1 &= -\frac{1}{3}\frac{\sin(v)(2\cos^2(v) - 1)}{d_2}, \\
b_2 &= \frac{1}{3}\frac{\sin(v)\left(16\cos^5(v) + 8\cos^4(v) - 16\cos^3(v) - 6\cos^2(v) + 4\cos(v) + 1\right)}{d_1}, \\
b_3 &= -\frac{1}{3}\frac{\sin(v)\left(8\cos^3(v) - 2\cos(v) + 1\right)\left(4\cos^3(v) - 4\cos(v) - 1\right)}{d_1}, \\
b_4 &= \frac{1}{6}\frac{\sin(v)\left(16\cos^4(v) + 24\cos^3(v) + 4\cos^2(v) - 2\cos(v) + 1\right)}{d_2}, \\
b_5 &= \frac{2}{3}\frac{\sin(v)\cos^2(v)(4\cos(v) + 3)}{d_1},
\end{aligned} \tag{67}$$

where

$$d_1 = v\cos(v)\left(8\cos^3(v) + 8\cos^2(v) - 1\right),$$

and

$$d_2 = v\cos(v)\left(4\cos^2(v) + 2\cos(v) - 1\right).$$

Table 6. The L_2 norm of the error for the fourth example using two implicit methods of trigonometric orders 2 and 3 and one explicit from [51] and one based on Runge-Kutta-Nyström.

Method	L_2 Error
AI2	0.312117(-14)
AI3	0.407362(-14)
(32)	0.470327(-18)
(66)	0.177952(-17)
(51)	0.218575(-37)
(46)	0.207559(-10)

Remark 5.

1. Adams implicit using the Taylor series for the coefficients did not improve the accuracy by using a higher order.
2. GMS of second trigonometric order gave better results than the Adams implicit. There is no improvement by using $q = 3$.
3. The method (51) is superior followed by GMS with $q = 2$ given by (32).
4. The method based on Runge-Kutta-Nyström could not compete with the others.

Example 5. *The fifth example is the "almost periodic" problem studied by Stiefel and Bettis [30]*

$$z'' + z = 0.001 e^{it}, \qquad 0 \le t \le 12\pi \tag{68}$$

subject to the initial conditions
$$z(0) = 1,$$
$$z'(0) = 0.9995i. \tag{69}$$

The exact solution is
$$z_{exact}(t) = \cos t + 0.0005t \sin t + i(\sin t - 0.0005t \cos t). \tag{70}$$

The solution represents motion on a perturbation of a circular orbit in the complex plane; the point $z(t)$ spirals slowly outwards.

The first order system equivalent was solved numerically using the above six methods. The results for $h = \pi/60$ and the exact value of $\omega = 1$ are given in Table 7.

Table 7. The L_2 norm of the error for the fifth example using the six methods of the previous example for the exact frequency.

Method	L_2 Error
AI2	0.446246(-9)
AI3	0.755130(-13)
(32)	0.777595(-12)
(66)	0.610169(-14)
(51)	0.693938(-38)
(46)	0.991561(-14)

It is clear that the methods of trigonometric order 3 are better than the lower order ones. Also the GMS is better than Adams implicit due to Gautschi [1]. Again, the method (51) is superior.

The next two examples demonstrate the quality of method for long-term integration.

Example 6. *The sixth example is the cubic oscillator as given in [52]*
$$y''(x) + y(x) = \epsilon y(x)^3, \qquad \epsilon = 10^{-3}, \tag{71}$$

with the initial conditions
$$y(0) = 1,$$
$$y'(0) = 0, \tag{72}$$

and the frequency $\omega = \sqrt{1 - 0.75\epsilon}$. The exact solution to cubic order in ϵ is given in [52]
$$y(x) = \cos(\omega x) + \frac{\epsilon}{128}(\cos(3\omega x) + \cos(\omega x)) + O\left(\epsilon^3\right).$$

The results are given in Table 8. It is clear that the methods that converged gave similar results. The methods (32) and (66) did not converge. The error is computed at $x = 2000\pi$.

Table 8. The L_2 norm of the error for the sixth example using the six methods of the previous example for the exact frequency.

Method	L_2 Error
AI2	0.209074(-3)
AI3	0.195944(-3)
(32)	Div
(66)	Div
(51)	0.150279(-3)
(46)	0.148082(-3)

Example 7. *The last example is a system of two second order ODEs describing two coupled oscillators with different frequencies, see [52].*

$$x''(t) + x(t) = 2\epsilon x(t)y(t),$$
$$y''(t) + 2y(t) = \epsilon x(t)^2 + 4\epsilon y(t)^3,$$
(73)

with initial conditions

$$x(0) = 1,$$
$$x'(0) = 0,$$
$$y(0) = 1,$$
$$y'(0) = 0,$$
(74)

where $\epsilon = 10^{-3}$. The frequencies $\omega_x = 1$ and $\omega_y = \sqrt{2} - \dfrac{3\epsilon}{2\sqrt{2}}$ can be found in [52]. We have compared the solution using the same methods to RKF45 of Maple. The L_2 norm of the difference between the solution of RKF45 and the six methods is given in Table 9. Now Adams implicit based method of trigonometric order 3 and our method (51) performed better than the others. Again, the methods due to Neta and Ford diverged.

Table 9. The L_2 norm of the error for the seventh example using the six methods of the previous example for the exact frequency.

Method	L_2 Error
AI2	0.401275
AI3	0.748492(-2)
(32)	Div
(66)	Div
(51)	0.552675(-2)
(46)	0.545854(-1)

7. Conclusions

We reviewed various trigonometrically-fitted methods and implemented representatives on several examples of second order initial value problems. In most examples our method from [51] was superior to others except for the first example for which Gautschi's method performed better. The methods (32) and (66) due to Neta and Ford failed to converge for the last two examples and in the long-term integration of example 1. The method based on Runge-Kutta-Nyström due to Demba et al. [49] could not compete with our method based on Obrechkoff.

Author Contributions: Conceptualization, B.N.; Investigation, B.N.; Software, C.C.; Writing—original draft, B.N.; Writing—review & editing, C.C.

Funding: This research was supported by Basic Science Research Program through the National Research Foundation of Korea (NRF) funded by the Ministry of Education (NRF-2016R1D1A1A09917373).

Acknowledgments: The first author thanks the Applied Mathematics Department at the Naval Postgraduate School for hosting him during the years. The authors thank the referees for their valuable comments.

Conflicts of Interest: The authors declare no conflict of interest.

References

1. Gautschi, W. Numerical integration of ordinary differential equations based on trigonometric polynomials. *Numer. Math.* **1961**, *3*, 381–397. [CrossRef]
2. Neta, B.; Ford, C.H. Families of methods for ordinary differential equations based on trigonometric polynomials. *J. Comput. Appl. Math.* **1984**, *10*, 33–38. [CrossRef]
3. Lambert, J.D. *Numerical Methods for Ordinary Differential Systems: The Initial Value Problem*; John Wiley & Sons Ltd.: Chichester, UK, 1991.

4. Chawla, M.M.; Neta, B. Families of two-step fourth order P-stable methods for second order differential equations. *J. Comput. Appl. Math.* **1986**, *15*, 213–223. [CrossRef]
5. Lambert, J.D.; Watson, I.A. Symmetric multistep methods for periodic initial value problems. *J. Inst. Math. Appl.* **1976**, *18*, 189–202 [CrossRef]
6. Brusa, L.; Nigro, L. A one-step method for the direct integration of structural dynamic equations. *Internat. J. Numer. Meth. Engng.* **1980**, *15*, 685–699. [CrossRef]
7. Van der Houwen, P.J.; Sommeijer, B.P. Explicit Runge-Kutta-Nyström methods with reduced phase errors for computing oscillating solutions. *SIAM J. Numer. Anal.* **1987**, *24*, 595–617. [CrossRef]
8. Thomas, R.M. Phase properties of high order, almost P-stable formulae. *BIT* **1984**, *24*, 225–238. [CrossRef]
9. Twizell, E.H. Phase-lag analysis for a family of two-step methods for second order periodic initial value problems. *J. Comput. Appl. Math.* **1986**, *15*, 261–263. [CrossRef]
10. Raptis, A.D.; Simos, T.E. A four-step phase-fitted method for the numerical integration of second order initial-value problems. *BIT* **1991**, *31*, 160–168. [CrossRef]
11. Simos, T.E. A two-step method with phase-lag of order infinity for the numerical integration of second order periodic initial-value problem. *Int. J. Comput. Math.* **1991**, *39*, 135–140. [CrossRef]
12. Simos, T.E. Explicit two-step methods with minimal phase-lag for the numerical integration of special second-order initial-value problems and their application to the one-dimensional Schrödinger equation. *J. Comput. Appl. Math.* **1992**, *39*, 89–94. [CrossRef]
13. Simos, T.E.; Raptis, A.D. Numerov-type methods with minimal phase-lag for the numerical integration of the one-dimensional Schrödinger equation. *Computing* **1990**, *45*, 175–181. [CrossRef]
14. Simos, T.E.; Vigo-Aguiar, J. A symmetric high order method with minimal phase-lag for the numerical solution of the Schrödinger equation. *Int. J. Mod. Phys. C* **2001**, *12*, 1035–1042. [CrossRef]
15. Obrechkoff, N. On mechanical quadrature (Bulgarian, French summary). *Spis. Bulgar. Akad. Nauk* **1942**, *65*, 191–289.
16. Achar, S.D. Symmetric multistep Obrechkoff methods with zero phase-lag for periodic initial value problems of second order differential equations. *Appl. Math. Comput.* **2011**, *218*, 2237–2248. [CrossRef]
17. Lambert, J.D.; Mitchell, A.R. On the solution of $y' = f(x,y)$ by a class of high accuracy difference formulae of low order. *Z. Angew. Math. Phys.* **1962**, *13*, 223–232. [CrossRef]
18. Ananthakrishnaiah, U. P-stable Obrechkoff methods with minimal phase-lag for periodic initial-value problems. *Math. Comput.* **1987**, *49*, 553–559. [CrossRef]
19. Van Daele, M.; Vanden Berghe, G. P-stable Obrechkoff methods of arbitrary order for second-order differential equations. *Numer. Algor.* **2007**, *44*, 115–131. [CrossRef]
20. Franco, J.M.; Rández, L. Explicit exponentially fitted two-step hybrid methods of high order for second-order oscillatory IVPs. *Appl. Math. Comput.* **2016**, *273*, 493–505. [CrossRef]
21. Vigo-Aguiar, J.; Ramos, H. A family of A-stable Runge-Kutta collocation methods of higher order for initial value problems. *IMA J. Numer. Anal.* **2007**, *27*, 798–817. [CrossRef]
22. Greenwood, R.E. Numerical integration of linear sums of exponential functions. *Ann. Math. Stat.* **1949**, *20*, 608–611. [CrossRef]
23. Brock, P.; Murray, F.J. The use of exponential sums in step by step integration. *Math. Tables Aids Comput.* **1952**, *6*, 63–78. [CrossRef]
24. Dennis, S.C.R. The numerical integration of ordinary differential equations possessing exponential type solutions. *Math. Proc. Camb. Phil. Soc.* **1960**, *56*, 240–246. [CrossRef]
25. Salzer, H.E. Trigonometric interpolation and predictor-corrector formulas for numerical integration. *ZAMM* **1962**, *42*, 403–412. [CrossRef]
26. Vigo-Aguiar, J.; Ramos, H. On the choice of the frequency in trigonometrically-fitted methods for periodic problems. *J. Comput. Appl. Math.* **2015**, *277*, 94–105. [CrossRef]
27. Van der Houwen, P.J.; Sommeijer, B.P. Predictor-corrector methods for periodic second-order initial-value problems. *IMA J. Numer. Anal.* **1987**, *7*, 407–422. [CrossRef]
28. Neta, B. Special methods for problems whose oscillatory solution is damped. *Appl. Math. Comput.* **1989**, *31*, 161–169.
29. Raptis, A.D.; Allison, A.C. Exponential-fitting methods for the numerical solution of the Schrödinger equation. *Comput. Phys. Commun.* **1978**, *14*, 1–5. [CrossRef]
30. Stiefel, E.; Bettis, D.G. Stabilization of Cowell's methods. *Numer. Math.* **1969**, *13*, 154–175. [CrossRef]

31. Cowell, P.H. *Essay on the Return of Halley's Comet*; In kommission bei W. Engelmann: Leipzig, Germany, 1910.
32. Quinlan, G.D.; Termaine, S. Symmetric multistep methods for the numerical integration of planetary orbits. *Astronom. J.* **1990**, *100*, 1694–1700. [CrossRef]
33. Simos, T.E.; Vigo-Aguiar, J. An exponentially-fitted high order method for long-term integration of periodic initial-value problems. *Comput. Phys. Comm.* **2001**, *140*, 358–365. [CrossRef]
34. Neta, B; Lipowski, Y. A New Scheme for Trajectory Propagation. *J. Astro. Sci.* **2002**, *50*, 255–268.
35. Vigo-Aguiar, J.; Ferrándiz, J.M. A general procedure for the adaptation of multistep algorithms to the integration of oscillatory problems. *SIAM J. Numer. Anal.* **1998**, *35*, 1684–1708. [CrossRef]
36. Vigo-Aguiar, J.; Martin-Vaquero, J.; Ramos, H. Exponential fitting BDF-Runge-Kutta algorithms. *Comput. Phys. Commun.* **2008**, *178*, 15–34. [CrossRef]
37. Martín-Vaquero, J.; Vigo-Aguiar, J. Exponential fitting BDF algorithms: Explicit and implicit 0-stable methods. *J. Comput. Math. Anal.* **2006**, *192*, 100–113. [CrossRef]
38. Neta, B. Families of Backward Differentiation Methods based on Trigonometric Polynomials. *Int. J. Comput. Math.* **1986**, *20*, 67–75. [CrossRef]
39. Sommeijer, B.P.; van der Houwen, P.J.; Neta, B. Symmetric linear multistep methods for second-order differential equations with periodic solutions. *Appl. Numer. Math.* **1986**, *2*, 69–77. [CrossRef]
40. Neta, B. Trajectory propagation using information of periodicity. In Proceedings of the AIAA/AAS Astrodynamics Specialist Conference, Boston, MA, USA, 10–12 August 1998; pp. 98–4577.
41. Simos, T.E. A P-stable complete in phase Obrechkoff trigonometric fitted method for periodic initial-value problems. *Proc. R. Soc. Lond. A* **1993**, *441*, 283–289. [CrossRef]
42. Wang, Z.; Zhao, D.; Dai, Y.; Wu, D. An improved trigonometrically fitted P-stable Obrechkoff method for periodic initial-value problems. *Proc. Math. Phys. Eng. Sci.* **2005**, *461*, 1639–1658. [CrossRef]
43. Neta, B. P-stable high-order super-implicit and Obrechkoff methods for periodic initial value problems. *Comput. Math. Appl.* **2007**, *54*, 117–126. [CrossRef]
44. Vanden Berghe, G.; Van Daele, M. Exponentially-fitted Obrechkoff methods for second-order differential equations. *Appl. Numer. Math.* **2009**, *59*, 815–829. [CrossRef]
45. Franco, J.M. A class of explicit two-step hybrid methods for second-order IVPs. *J. Comput. Appl. Math.* **2006**, *187*, 41–57. [CrossRef]
46. Chawla, M.M. Numerov made explicit has better stability. *BIT* **1984**, *24*, 117–118. [CrossRef]
47. Tsitouras, C. Explicit Numerov type methods with reduced number of stages. *Comput. Math. Appl.* **2003**, *45*, 37–42. [CrossRef]
48. Fang, Y.; Song, Y.; Wu, X. Trigonometrically fitted explicit Numerov-type method for periodic IVPs with two frequencies. *Comput. Phys. Commun.* **2008**, *179*, 801–811. [CrossRef]
49. Demba, M.A.; Senu, N.; Ismail, F. New explicit trigonometrically-fitted fourth-order and fifth-order Runge-Kutta-Nyström methods for periodic initial value problems. *Int. J. Pure Appl. Math.* **2016**, *109*, 207–222. [CrossRef]
50. Vanden Berghe, G.; Van Daele, M. Trigonometric polynomial or exponential fitting approach? *J. Comput. Appl. Math.* **2009**, *233*, 969–979. [CrossRef]
51. Chun, C.; Neta, B. A new trigonometrically-fitted method for second order initial value problems. **2019**, in preparation.
52. Vigo-Aguiar, J.; Simos, T.E.; Ferrándiz, J.M. Controlling the error growth in long-term numerical integration of perturbed oscillations in one or several frequencies. *Proc. Math. Phys. Eng. Sci.* **2004**, *460*, 561–567. [CrossRef]

© 2019 by the authors. Licensee MDPI, Basel, Switzerland. This article is an open access article distributed under the terms and conditions of the Creative Commons Attribution (CC BY) license (http://creativecommons.org/licenses/by/4.0/).

MDPI
St. Alban-Anlage 66
4052 Basel
Switzerland
Tel. +41 61 683 77 34
Fax +41 61 302 89 18
www.mdpi.com

Mathematics Editorial Office
E-mail: mathematics@mdpi.com
www.mdpi.com/journal/mathematics

www.ingramcontent.com/pod-product-compliance
Lightning Source LLC
LaVergne TN
LVHW070637100526
838202LV00012B/827